The Holy Pattern

The Holy Pattern

How the Universe Organizes Meaning

Elodie Varkash

Mindful Pages

Published in 2026

ISBN: 9789375368540 (PB)
ISBN: 9789375360636 (eBook)

Published by

Mindful Pages
Imprint of Alpha Editions LLC
312 W. 2nd St #1834
Casper, WY 82601, USA

This book was developed using advanced writing tools to support the drafting and ideation process. All content has been thoughtfully curated, refined, and reviewed to ensure quality and relevance.

Content

Introduction

The floor beneath your feet is tiled. You do not notice it at first. Your attention is elsewhere: the conversation you are replaying, the errand you are late for, the weight of the day already forming on your shoulders. Then, for no obvious reason, your gaze drops. The tiles repeat—square after square, light and dark in quiet alternation. The pattern has been there all along, holding your steps without demanding thanks. You pause for a moment, feel a strange flicker of recognition, and move on. Something has brushed past your awareness, unnamed.

Later, perhaps on a walk, you notice branches above a footpath crossing and recrossing like deliberate strokes. Or you hear a chant repeated in a place of worship, the same phrase returning until it loosens something in you. The experience is modest, almost forgettable. Yet the forms linger. You begin to sense that the world is speaking in repetitions, and that you are only half-listening.

There is a quiet paradox here. The same shapes that decorate a tiled floor also govern the growth of shells and galaxies. The same ratios that steady a piece of music recur in architecture, prayer, and the physics of waves. Spirals, lattices, symmetries, cycles—these are not rare curiosities but the most common handwriting of reality. We pass them daily without comment. This book begins from the suspicion that such familiarity is precisely the problem.

Pattern is usually treated as background. It is what remains once the "real" content has been extracted: the melody beneath the lyrics, the design around the doctrine, the geometry supporting the building in which belief is enacted. Modern habits of thought have trained us to see pattern as either decoration or mechanism. At best, it is admired. At worst, it is ignored. Rarely is it addressed as meaning in its own right.

Yet human beings have not always been so inattentive. For most of our history, pattern was a primary way of knowing. It ordered calendars and rituals, shaped laws and myths, disciplined bodies and communities. To trace a circle, to mark a threshold, to repeat a word at fixed intervals—these were not aesthetic preferences but acts of interpretation. They were claims about how the world is arranged and where a human life might fit within that arrangement.

The central question of this book is simple to state and difficult to evade. What if pattern is not merely something the universe happens to display, but something it is doing? What if repetition is not an accident of matter, but a mode of expression? To ask this is not to smuggle superstition into science or to replace analysis with mysticism. It is to take seriously a fact that cuts across disciplines and centuries: wherever meaning is found, pattern is already there, organising it.

This question matters now for reasons that are both urgent and ancient. We live surrounded by unprecedented quantities of data, yet struggle to agree on what any of it means. Algorithms sort our attention, forecasts shape our fears, and invisible systems nudge behaviour with mathematical confidence. Pattern has become powerful again, but largely unexamined. At the same time, many inherited forms of meaning—religious, communal, ethical—have lost their persuasive force or hardened into brittle formulas. We are hungry for order, yet suspicious of it. We crave coherence, yet fear manipulation. The result is a cultural tension that expresses itself everywhere, from personal anxiety to political fracture.

This tension is not new. Ancient observers faced a sky they could not control and a landscape that demanded interpretation. They watched the seasons turn, the stars return, the tides obey hidden rhythms. From these observations came calendars, myths, and temples; from those came shared ways of life. Pattern was not imposed on reality but drawn from it, slowly and with care. Errors were made, of course, and abuses followed. Yet the underlying intuition—that the world's regularities might teach us how to live—was neither naïve nor arbitrary.

Modern science has complicated this picture without dissolving it. The more deeply we probe the structure of matter and energy, the more insistently pattern appears. Atoms arrange themselves in crystalline lattices. Rivers branch according to mathematical constraints. Biological systems repeat successful forms across scales. Even processes that look chaotic reveal statistical order when viewed patiently enough. The universe, it turns out, is not a formless flux interrupted by occasional regularity. It is patterned through and through.

This book does not argue that every pattern is meaningful, nor that meaning can be reduced to geometry. It argues something more demanding. Pattern is a necessary condition for meaning to arise at all. Without repetition, there can be no recognition. Without proportion, no sense of balance. Without rhythm, no shared time. Meaning is not an extra layer pasted onto reality by human minds; it is a relationship between minds and the structured world they inhabit.

To make this case, the chapters that follow move deliberately across domains that are often kept apart. Geometry and worship, neuroscience and ritual, architecture and ethics, fractals and personal growth—these are treated not as metaphors for one another but as overlapping expressions of a single concern. The concern is organisation: how order appears, how it is perceived, how it guides action, and how it sometimes deceives.

A recurring motif will accompany us throughout. Think of pattern as grammar. Grammar does not dictate what must be said, but it makes saying anything intelligible. It constrains and enables at once. It allows creativity without collapse. In the same way, the patterns of the universe form a grammar of existence. They do not tell us what to believe or how to act in detail, but they shape the space in which belief and action make sense. To learn this grammar is not to memorise rules; it is to become attentive to form.

Attention, in fact, is the quiet discipline at the heart of this book. Learning to see pattern is not a technical skill so much as a shift in perception. Many people report a moment when this shift occurs: suddenly spirals appear everywhere, symmetries announce themselves, repetitions refuse to be ignored. What begins as fascination can become disorientation. If the world is so ordered, what does that imply about freedom? If patterns repeat, where does novelty enter? These questions are not obstacles to be cleared away but invitations to think more carefully about what order actually is.

Order, as we shall see, is not the opposite of change. It is what makes change legible. Fractals show us how complexity can arise from simple rules. Rhythms teach us how time can be structured without being frozen. Rituals reveal how repetition can deepen meaning rather than exhaust it. At every turn, the book resists the false choice between rigid structure and chaotic freedom. The universe is more subtle than that, and so are we.

The stakes are not merely intellectual. How we interpret pattern shapes how we live. When patterns are treated as absolute, they become oppressive. When they are dismissed as illusory, life fragments. Between superstition and cynicism lies discernment: the patient work of telling which regularities deserve trust and which must be questioned. This is as true in spiritual life as it is in politics, economics, and personal relationships.

The chapters on false orders address this directly. Human beings are exquisitely skilled at finding patterns, and that skill can betray us. Coincidences harden into convictions. Algorithms masquerade as oracles. Conspiracies offer the comfort of hidden coherence without the discipline of evidence. To speak of holy pattern in such a context demands humility. Not every order is sacred. Not every repetition reveals truth. The ability to say "this pattern misleads" is as important as the ability to say "this pattern teaches".

Yet scepticism alone cannot sustain a culture. It can expose error but not generate meaning. For that, we must return to practices that train perception rather than inflame it. Walking a labyrinth. Listening to chant. Observing the body's own rhythms. Designing spaces that invite attention rather than scatter it. These are not escapes from reality but engagements with its structure at a human scale.

Throughout the book, nature is treated not as a sentimental backdrop but as a primary source of instruction. Spirals in shells, cycles in seasons, symmetries in snowflakes—these are not proofs of doctrine, but they are reminders that constraint and beauty are not enemies. Matter itself seems inclined toward form. To notice this is not to deny randomness or suffering; it is to refuse the claim that disorder has the final word.

Music, poetry, and architecture enter the discussion not as cultural luxuries but as laboratories of pattern. They show how number becomes experience, how repetition becomes memory, how space and time can be shaped to hold communities together. A cathedral and a calendar perform related work. Both impose structure on human attention. Both can be abused. Both can also, at their best, teach a people how to inhabit time and space with care.

The later chapters turn explicitly toward ethics and the future. If pattern organises meaning, then moral life cannot be reduced to isolated choices. It has shape. Balance, reciprocity, proportion—these are not vague ideals but spatial concepts that have guided justice systems for millennia. To imagine ethics as geometry is to ask different questions: Where is the centre? What has been excluded? Which relationships are distorted? Such questions do not guarantee agreement, but they sharpen responsibility.

New technologies make these questions unavoidable. Digital environments are designed spaces with consequences for attention, belief, and behaviour. Their patterns are intentional, often opaque, and rarely accountable. To navigate them well requires more than technical literacy; it requires a trained sense of pattern's power. Without that, we remain at the mercy of forms we did not choose.

This book offers no final map of the universe. It does not promise certainty where none is possible. It offers something more modest and more demanding: a way of seeing that treats pattern as a serious partner in the search for meaning. It asks the reader to slow down, to look again at what repeats, to consider why certain forms endure across cultures and epochs.

As you read, questions will arise that resist quick answers. Is the universe ordered for us, or are we ordered for it? Does recognising pattern deepen freedom or limit it? Can a life be consciously shaped without becoming rigid? Where does reverence belong in a world explained by equations? These questions are not puzzles to be solved and set aside. They are pressures that, properly borne, clarify what matters.

The chapters that follow trace a path through seeing, understanding, discerning, and living by pattern. They move from perception to practice, from observation to responsibility. Along the way, familiar objects—floors, songs, seasons, bodies—are asked to speak more clearly than we usually allow. They have been repeating themselves for a long time.

The invitation is to listen.

Chapter 1
Learning to See the Holy Pattern

The first time it happens, it often feels trivial. Someone notices the curve of a fern unfurling and realises it resembles the spiral of a shell they once held as a child. Or they see the same triangular tension appear in a bridge, a painted icon, and the posture of three people in conversation. There is a flicker of pleasure, a sense of cleverness perhaps, and then an unease. The repetition refuses to stay put. Once noticed, it multiplies. The world begins to echo itself.

Many people describe this as a moment of awakening, though that word risks melodrama. It is quieter than that. More like learning to hear a harmony that had always been present beneath the melody. Streets suddenly appear gridded rather than chaotic. Leaves disclose branching laws. Even the spaces between things—gaps in fences, pauses in music, intervals in speech—seem to obey rules. What was once background becomes insistently foregrounded.

The surprise lies not in the existence of pattern, but in its abundance. No one is shocked to find order in a mosaic or a piece of music; that is what they are for. The shock comes when the same structures appear where no designer seems to be at work: in clouds, in crowds, in the rise and fall of attention over a day. Pattern recognition is often presented as a specialist skill, the province of mathematicians or artists. In practice, it is a destabilising expansion of ordinary perception. It asks a question that cannot be unasked: if the same forms recur across domains, what connects them?

This chapter begins with that shock of recognition and treats it seriously, not as a trick of the eye but as a meaningful cognitive event. To see pattern is not simply to notice similarity; it is to intuit relationship. The experience carries both exhilaration and cost. Some people report delight at the sudden coherence of things. Others feel unsettled, as if the world has grown less accidental than they were taught to expect. The gift arrives with a disruption attached.

From there, the chapter moves deliberately into mess. Clouds are an obvious place to start. They refuse edges, defy categorisation, and shift too quickly to be captured. Yet meteorologists speak of systems, cells, and fronts. Satellite images reveal repeating vortices and bands. What looks shapeless from the ground exhibits structure from above. The same is true of markets, whose daily fluctuations appear irrational until patterns of boom, panic, and correction are traced across decades. City streets, too, seem unruly until traffic engineers map flows, bottlenecks, and emergent behaviours that no single driver intends.

These examples matter because they unsettle a comfortable assumption: that order must be imposed from outside. In many cases, order emerges from interaction. Birds flock without a leader, following a handful of simple rules about distance and alignment. Traffic jams propagate backwards along a motorway without any accident at their source. The beauty here is not ornamental; it is functional. Structure arises because it works. Efficiency, not aesthetics, is the driving force, and yet the result is often something we are tempted to call beautiful.

The temptation deserves examination. Why should efficiency look beautiful to us at all? Why should repeated forms soothe the eye or steady the breath? The answer lies partly in our bodies. Human nervous systems evolved in patterned environments: day and night, seasons, heartbeats, footsteps. Regularity signals predictability, and predictability lowers threat. Rituals draw on this biology. A repeated prayer does not merely convey words; it entrains attention. Familiar designs stabilise communities because they stabilise expectation. People know where to stand, when to speak, how to move. Anxiety recedes.

Yet the same mechanisms can suffocate. When patterns become rigid, they cease to guide and begin to constrain. A ritual performed without reflection can hollow out meaning. A social structure repeated beyond its usefulness can trap rather than protect. The human response to pattern is therefore ambivalent. We long for it and resist it. We seek repetition for comfort and novelty for freedom, often without noticing how tightly the two are intertwined.

This ambivalence is not a modern invention. Religious traditions have wrestled with it for millennia. Sacred texts warn against empty repetition even as they prescribe it. Architectural canons insist on proportion while allowing endless variation. The tension is productive. It forces communities to ask whether a pattern is serving life or merely preserving itself. Learning to see pattern clearly includes learning to ask that question without flinching.

The chapter's central claim is that pattern has never been merely decorative. To treat it as such is a recent habit, born of a split between form and meaning that has proved difficult to sustain. In illuminated manuscripts, the placement of a knot or a spiral was not an afterthought. It carried doctrinal weight. The interlacing lines on a textile encoded cosmology, genealogy, and moral order. A building's geometry taught theology long before its sermons were heard.

Consider the repetition of arches in a cloister. Each arch is identical, yet walking beneath them is not monotonous. The body moves through a rhythm of shadow and light, compression and release. Time is measured in steps. The pattern instructs without speaking. It shapes patience. To call this decoration is to misunderstand its function. It is pedagogy rendered in stone.

Learning to read such patterns requires a shift in posture. The modern eye has been trained to skim surfaces for information and move on. Pattern asks for lingering. It rewards attention rather than speed. This chapter does not offer a catalogue of symbols to memorise; it offers a way of looking that can be practised anywhere. The tiled floor returns, but now as a text. The chant repeats, but now as a form of thinking. The city becomes legible without becoming predictable.

There is a risk here, and it must be named early. Pattern recognition can tip into obsession. Seeing order everywhere can flatten difference and invent connections that do not exist. This chapter is not an invitation to abandon judgement. On the contrary, it insists that discernment begins with careful seeing. The difference between a helpful pattern and a harmful one often lies not in its complexity but in how it responds to reality. Does it adapt, or does it demand compliance?

Modern relevance presses in at this point. Digital environments are saturated with patterns designed to capture attention. Feeds refresh at intervals tuned to human impatience. Notifications arrive in rhythms calibrated to keep the nervous system slightly off-balance. These are not neutral designs. They exploit the same cognitive machinery that makes ritual effective. To live well among such forces requires an education in pattern that is both appreciative and critical.

The chapter therefore moves between wonder and warning. It celebrates the moment when the world's repetitions reveal themselves, and it insists on asking what those repetitions are doing to us. This is not a call to retreat into abstraction. It is a call to become more concrete. Pattern lives in bodies, buildings, and habits. It is encountered by walking, listening, and watching. Learning to see it changes how one inhabits space and time.

Throughout the chapter, a simple image recurs: the idea of tuning. Just as a musical instrument must be tuned to resonate properly, perception must be adjusted to detect pattern without distortion. Too loose, and everything blurs. Too tight, and nothing new can enter. The aim is not perfect pitch but responsive hearing. Patterns are not static objects; they are dynamic relationships. They unfold.

By the end of the chapter, the reader will not be handed a set of conclusions. Instead, they will have been introduced to a skill: the ability to notice repetition without reducing it, to appreciate order without absolutising it. They will have walked through examples that range from the everyday to the sacred, from the scientific to the psychological, all pointing toward a single insight. Pattern is already shaping our lives. The question is whether we are conscious participants or passive recipients.

The chapters that follow will deepen and complicate this insight. They will give names to shapes, trace their histories, and test their limits. But this first step is foundational. Without the willingness to see, the rest collapses into abstraction. With it, even the most familiar scenes begin to disclose unexpected depth.

The shock of recognition, once felt, does not fade. It matures. It becomes a steady attentiveness, a readiness to ask what is being repeated and why. This chapter opens that door. The invitation is to step through it slowly, eyes open, and allow the world's quiet grammar to begin its work.

The shock of recognition

In 1975, at IBM's Thomas J. Watson Research Center in Yorktown Heights, New York, Benoît Mandelbrot published *Les Objets Fractals*. The book's title was technical, almost forbidding, yet its effect was oddly intimate. Mandelbrot gathered images of coastlines, clouds, and financial price series and showed that the same jagged forms recur at different scales. Shorelines refused to smooth out under measurement; zooming in revealed more of the same. The finding was not merely mathematical. It altered how people saw the world. Landscapes once dismissed as irregular acquired a grammar. Disorder gained a family resemblance.

The reaction was not confined to laboratories. Artists, architects, and engineers began to notice repetitions they could not unsee. Walks along the coast turned into lessons in self-similarity. A fern became an argument. A market chart looked less arbitrary. Pattern announced itself, and then persisted. Seeing changed. Permanently.

It changes you.

The shock here is not the discovery of a new shape. It is the discovery of recurrence across boundaries that had seemed firm. Nature, art, and worship had been filed into different mental drawers. Suddenly the same ratios and curves appeared in all three. Spirals in shells echoed staircases and votive designs. Grids beneath cities matched the pages of sacred manuscripts. The experience is exhilarating, but it also carries a faint vertigo. If the same forms govern such different domains, what exactly have we been looking at all this time?

Psychologists have a name for the moment when perception reorganises itself. In 1896, the philosopher and psychologist William James described what he called a "habitual way of seeing" being interrupted, producing a sudden clarity that feels both obvious and new. The interruption is rarely neutral. It brings pleasure, but it also threatens settled categories. Pattern recognition, once triggered, does not politely restrict itself to approved contexts. It spills.

The gift is easy to describe. The disruption is harder.

Consider the way ratios migrate. In the sixth century BCE, Pythagorean thinkers in southern Italy noticed that simple numerical ratios governed musical intervals. An octave corresponded to a doubling of frequency; a fifth to a ratio of 3:2. These relationships were audible and measurable. What followed was a leap. If number could order sound so elegantly, perhaps it ordered the cosmos as well. Harmony became a property of reality, not merely of music. The idea travelled. By the twelfth century, builders of Romanesque churches across Europe were laying out naves and transepts according to proportional schemes that echoed musical consonance. Stone learned from sound.

Show. Name. Explain. Implication. The pattern is not decorative; it is structural. Ratios that stabilise hearing also stabilise buildings. The implication is unsettling. Proportion appears to be a bridge between sense and meaning. Once noticed, that bridge cannot be dismantled without loss.

A modern parallel appears in a domain far removed from chant and masonry. In 1968, traffic engineers in the United Kingdom documented a phenomenon on the M4 motorway near Slough: waves of congestion travelling backwards along the road at roughly 15 kilometres per hour, even in the absence of accidents or lane closures. The observation was later formalised by the physicist Dirk Helbing in the 1990s. Individual drivers behaved rationally; the system did not. The pattern emerged from interaction, not intention. What looked like chaos from behind the wheel revealed disciplined motion when measured.

The visible behaviour is simple. Drivers brake. Distances compress. A wave forms. The metric is speed variance over time. The decision implication is sobering. No amount of individual virtue eliminates the pattern. Only changes to spacing rules do. Order is not imposed; it arises.

Once again, recognition brings a double response. There is relief in knowing that the jam is not personal. There is unease in realising how little agency the individual possesses within the pattern. The motorway becomes a moving diagram, and the driver a point within it.

This ambivalence surfaces sharply in religious practice. Repetition lies at the heart of worship: fixed hours, recurring prayers, cycles of fasting and feast. Anthropologists have long noted the stabilising effect. In

1912, Émile Durkheim described ritual as a mechanism for synchronising emotion and belief within a group. The behaviour is observable. Bodies move together. Voices align. The metric is coherence: reduced variance in gesture and speech. Trust increases.

Yet repetition also attracts suspicion. In the Christian Gospels, Jesus warns against "vain repetitions" in prayer, a phrase later elaborated by theologians wary of mechanical piety. The counter-example matters. Pattern can hollow out intention. A rosary recited without attention becomes noise. The reconciliation lies not in abandoning repetition but in distinguishing between form that trains attention and form that replaces it. The shock of recognition here is ethical. The same structure can heal or harm, depending on how it is inhabited.

The motif threading these cases is simple: a curve drawn again and again. The curve is not always literal. Sometimes it is temporal, sometimes spatial, sometimes behavioural. But it returns. Each return sharpens the question. Are we discovering patterns, or are we projecting them?

Cognitive science complicates the picture. Humans are exceptional pattern detectors. In 1944, the psychologist B. F. Skinner demonstrated how pigeons, rewarded at random intervals, developed elaborate rituals—turning in circles, pecking corners—believing these actions caused the reward. The experiment, conducted at Harvard University, is often cited as evidence of superstition. The visible behaviour is ritualised movement. The metric is reward frequency. The implication is cautionary. Pattern recognition can misfire.

This limitation cannot be ignored. To see repetition everywhere is not the same as understanding it. Clouds form vortices because of fluid dynamics, not because they "mean" anything. Financial charts repeat shapes because of feedback loops, not fate. The shock of recognition becomes dangerous when it slides into overreach.

And yet, the dismissal of pattern as mere projection fails to explain its persistence across cultures and disciplines. Consider Islamic geometric art. In the ninth century, craftsmen in Baghdad developed complex tiling patterns based on repeated polygons and stars. The prohibition against figurative imagery forced attention onto form itself. The behaviour is measurable. Tiles repeat according to strict rules. The metric is symmetry group classification, later formalised by mathematicians. The implication is theological. Infinity is suggested through repetition without centre. God is not depicted; order is.

Here, pattern is not inferred; it is chosen. The recognition is intentional, pedagogical. Worshippers are trained to see beyond surface variety to underlying unity. The pattern does not point to itself. It points through.

The disruption arrives when such practices are placed alongside modern habits of attention. Screens refresh unpredictably. Notifications arrive at irregular intervals. The nervous system adapts. A 2019 report by the American

Psychological Association documented rising stress linked to fragmented attention, measured through self-reported anxiety and sleep disruption. The behaviour is constant checking. The metric is interruption frequency per hour. The implication is practical. Environments shape perception.

Against this backdrop, the sudden recognition of stable pattern can feel like relief. A tiled floor steadies the eye. A chant steadies the breath. The shock is not only cognitive; it is physiological. Regularity lowers arousal. Heart rate variability increases. Calm follows. The gift is embodied.

But the same relief exposes dependence. If calm requires repetition, what happens when repetition is withdrawn or manipulated? This is not a hypothetical question. Digital platforms design rhythms deliberately. The curve returns, this time as a usage graph.

The counter-example that guards against naïveté comes from modern art. In 1917, Marcel Duchamp submitted a porcelain urinal titled *Fountain* to the Society of Independent Artists in New York. It was rejected. The gesture was a refusal of inherited pattern. No proportion, no repetition, no reverent geometry. The outcome was catalytic. Art was forced to ask whether meaning required form at all. The limitation exposed is real. Pattern is not the only carrier of significance. Shock can also arise from rupture.

Yet even rupture becomes a pattern once repeated. The avant-garde hardens into a style. Negation acquires its own grammar. Duchamp's provocation did not abolish form; it relocated it. The reconciliation is instructive. Pattern is not a prison, but neither is it optional. It returns, whether welcomed or resisted.

What distinguishes the shock of recognition from a passing curiosity is its effect on conduct. People who experience it begin to compare. They measure. They notice when a space calms them and when it agitates them. They observe how repeated actions alter mood. The abstraction "meaning" translates into visible decisions: where to walk, how to pray, which rooms to inhabit longer. Behaviour shifts.

This is why the experience cannot be reduced to a niche hobby. It is not about collecting patterns as one collects stamps. It is about recalibrating attention. Once recalibrated, attention becomes harder to exploit and easier to educate. The curve, encountered again, becomes a diagnostic tool.

The shock does not resolve into certainty. It opens a tension that persists. If pattern is everywhere, discernment matters more, not less. Which repetitions deserve trust? Which ratios clarify, and which coerce? The question is not answered here. It is carried forward.

Look again at the curve. It is still there.

From chaos to hidden order

On a humid morning in May 2008, a team led by Yuki Sugiyama at Nagoya University gathered twenty-two cars on a circular test track in Japan. There were no intersections, no obstacles, no instructions beyond "drive normally." Within minutes, a traffic jam formed. It travelled backwards around the circle at a steady speed, though no driver caused it and none could prevent it. The experiment, later published in the *New Journal of Physics* (Sugiyama et al., 2008), produced a concrete outcome that startled even seasoned engineers: congestion emerged spontaneously from uniform motion. Order appeared inside apparent disorder.

Nothing had gone wrong. Everything behaved as expected.

That is the unsettling lesson. Systems we experience as messy, frustrating, or chaotic often obey rules that are both strict and indifferent to individual intention. The shock is not that structure exists, but that it operates precisely where we assume there is none. Once this becomes visible, the world looks less accidental and more exacting.

Chaos, in everyday speech, means confusion. In science, it names something narrower and more interesting: sensitivity to initial conditions. Small differences amplify. Outcomes diverge. The surface looks unruly, yet the system follows deterministic rules. Clouds form this way. So do markets. So do cities at rush hour. The hidden order is not tidy. It does not promise comfort. It promises legibility.

Watch a summer sky. Clouds swell, shear, dissolve. No two are alike. Yet satellite imagery reveals repeating bands, vortices, and cells governed by fluid dynamics. In 1963, the meteorologist Edward Lorenz demonstrated that tiny variations in atmospheric input could radically alter long-term weather patterns. His equations, developed at MIT, showed why prediction fails without denying structure. The visible behaviour is billowing, transient forms. The metric is sensitivity to initial conditions. The implication is restraint. Chaos does not mean randomness; it means bounded unpredictability.

Markets offer a harsher classroom. Price movements appear erratic, driven by rumour and fear. Yet long-term charts display recurring shapes: bubbles, crashes, recoveries. In 1973, Fischer Black and Myron Scholes published a model describing option pricing, assuming markets followed a form of random walk. The model gained practical traction, shaping trading floors in Chicago and London. The abstraction was risk. The visible behaviour was hedging frequency. The metric was volatility. The implication was confidence.

That confidence proved fragile. The 1987 stock market crash exposed how feedback loops could overwhelm assumptions of independence. Selling begat selling. Programmed responses amplified movement. What looked like collective panic was also mechanical synchrony. The hidden order did not prevent collapse; it explained its speed.

Cities repeat the lesson in concrete and steel. Streets teem with pedestrians whose paths intersect unpredictably. Stand at a busy crossing in Shibuya, Tokyo, and the scene seems anarchic. Yet researchers at the University of Tokyo, analysing pedestrian flow data in 2015, showed that people spontaneously form lanes, minimising collision without instruction. The behaviour is visible. Bodies align. The metric is collision avoidance rate. The implication is modest but profound. Coordination can emerge without command.

This emergence is not mystical. It rests on local rules: keep a certain distance, match speed, adjust direction. Birds flock the same way. In 1987, Craig Reynolds, then at Symbolics Inc. in Massachusetts, simulated flocking behaviour using three simple parameters: separation, alignment, cohesion. The resulting computer models displayed lifelike motion. No leader appeared. The pattern held. The image returns: a curve traced again and again.

Show. Name. Explain. Implication. The structure persists across scales.

The comfort this knowledge offers is limited. Knowing that traffic jams form predictably does not shorten the commute. Knowing that bubbles recur does not prevent loss. Hidden order is descriptive before it is redemptive. It explains why certain failures repeat. It does not guarantee their correction.

This limitation matters. Pattern recognition can seduce. If order exists beneath mess, perhaps it can be mastered. That hope fuels technocratic ambition. In the 1960s, urban planners inspired by systems theory redesigned neighbourhoods around traffic flow and zoning efficiency. The Pruitt–Igoe housing complex in St. Louis, completed in 1956, embodied such confidence. By the early 1970s, it was deemed uninhabitable and demolished. The visible behaviour was social breakdown. The metric was vacancy rate. The implication was chastening. Abstract order ignored lived complexity.

Here lies the counter-example that disciplines enthusiasm. Hidden order does not equate to optimal order. Systems can be coherent and cruel. Efficient flows can erase human scale. The reconciliation requires a distinction between describable patterns and desirable ones. The former are unavoidable. The latter demand judgement.

Return to clouds. Their beauty does not imply benevolence. Storm systems organise energy efficiently; they also destroy. Markets allocate resources; they also concentrate risk. Traffic flows maximise throughput; they also privilege speed over safety. The question shifts. Not "Is there order?" but "What does this order do?"

The quiet motif threading these domains is movement. Cars circle. Birds turn. Prices swing. Clouds roll. Movement reveals structure because it exposes constraint. Static snapshots deceive. Motion discloses rule.

This has consequences for how we read environments of belief. Rituals move bodies through space and time.

Processions follow fixed routes. Prayer hours punctuate days. From a distance, the repetition looks monotonous. From within, it generates orientation. The visible behaviour is coordinated motion. The metric is synchronisation. The implication is belonging.

Yet even here, hidden order can harden. When movement becomes compulsory rather than invitational, pattern turns coercive. History records such failures. In early twentieth-century Europe, mass rallies employed choreography and repetition to fuse identity and obedience. The order was exact. The outcome was catastrophic. The lesson is severe. Discipline without discernment is dangerous.

Modern technologies intensify the stakes. Algorithmic systems detect patterns in behaviour at scales no human observer can match. Recommendation engines compare, cluster, predict. The abstraction is preference. The visible behaviour is click-through rate. The metric is dwell time. The implication is influence. Order is extracted and fed back, shaping what appears next.

Here, the shock of hidden order arrives late. Users experience feeds as chaotic, endless, personalised. The structure remains unseen until outcomes accumulate: polarisation, addiction, fatigue. Once recognised, the order cannot be unrecognised. The decision then looms. Accept the pattern, resist it, or redesign it?

The chapter does not resolve this. It insists on clarity first. Hidden order is neither enemy nor saviour. It is a condition of complex systems. To deny it invites superstition; to worship it invites tyranny. The work lies between.

Measure again. Compare again. Commit carefully.

The final tension remains. When faced with apparent chaos, do we search for the rule beneath it, or do we ask whether the rule deserves to endure?

Why pattern comforts and unsettles

On 24 August 1662, in London, the Act of Uniformity came into force under the authority of the Parliament of England. Clergy were required to use the newly revised *Book of Common Prayer* exactly as prescribed. About two thousand ministers refused and were ejected from their livings in what became known as the Great Ejection. The outcome was immediate and measurable: parishes lost familiar pastors overnight; congregations fractured; dissenting traditions hardened. A single pattern of worship, imposed for the sake of unity, delivered both order and exile.

The book itself was not new. Nor was its cadence. Its prayers had steadied English worship for a century. What changed was enforcement. The same repetitions that had comforted generations became instruments of exclusion. Pattern proved double-edged.

Count the beats.

The comfort of pattern begins in the body. Repetition stabilises. Breath settles into rhythm. Muscles learn what comes next. Anthropologists observing ritual across cultures have repeatedly noted the same visible behaviours: synchronised movement, shared vocal tempo, predictable sequences. In 1914, Bronisław Malinowski described Trobriand Island fishing rites in which repeated chants preceded dangerous voyages. The behaviour was consistent. The metric was participation rate. The implication was trust. When the sea could not be controlled, the sequence could.

Trust here is not abstract. It shows itself in posture and timing. People arrive on time. They stand together. They speak together. The metronome ticks. In Benedictine monasteries following the Rule of St Benedict, written in Italy in the sixth century, the Divine Office structures the day into fixed hours of prayer. The decision to stop work at set times produces a visible outcome: labour yields to listening. The metric is interruption frequency. The implication is humility. No task outranks the bell.

Such patterns do more than organise time. They reduce cognitive load. Neuroscientific studies of rhythm, including work by Aniruddh Patel published in 2008, show that predictable temporal structures allow the brain to allocate attention more efficiently. The behaviour is reduced variance in response time. The metric is synchrony. The implication is calm. When the next beat is known, vigilance relaxes.

This is why repeated prayers endure. The words do not need to be invented. Attention can move elsewhere. Meaning deepens through familiarity rather than novelty.

In the Taizé Community in eastern France, founded in 1940 by Roger Schutz, chants built on short, repeated phrases are sung by thousands of pilgrims each year. The outcome is observable. Voices blend despite linguistic difference. The metric is sustained participation. The implication is inclusion. Pattern carries what speech cannot.

Yet the same mechanisms that soothe can constrain. Rigidity enters when repetition ceases to be chosen. The Great Ejection remains instructive not because uniform worship is inherently oppressive, but because enforcement transformed form into boundary. The visible behaviour shifted. Clergy left pulpits. Congregations split. The metric was attendance loss. The implication was fracture.

This tension recurs wherever pattern migrates from invitation to requirement. In 1890s Japan, as the Meiji government formalised State Shinto, rituals once embedded in local practice were standardised and tied to national identity. Participation became compulsory. The behaviour was attendance at shrine rites. The metric was compliance. The implication was politicisation. What had stabilised communities now disciplined them.

The counter-example sharpens the point. Patterns that retain elasticity tend to endure without coercion. Jewish liturgical practice offers one such case. The Amidah prayer, central to daily worship, follows a fixed structure established by the early rabbinic period. Yet its content has been adapted across centuries and communities. The visible behaviour is standing prayer. The metric is structural consistency with textual variation. The implication is continuity without stagnation.

Elasticity matters because human lives change. Patterns that cannot accommodate grief, dissent, or discovery become brittle. When they crack, they injure. The psychology of this is not obscure. In 1971, the psychologist Martin Seligman described learned helplessness in experiments where subjects exposed to unavoidable shocks ceased attempting escape. Translate the abstraction. When individuals experience rigid sequences without agency, behaviour narrows. The metric is reduced exploratory action. The implication is despair.

Spiritual lives are not exempt. Repeated forms that once carried meaning can, under pressure, become hollow. The behaviour is rote recitation without engagement. The metric is attentional drift. The implication is alienation. This is not a failure of pattern itself, but of responsiveness.

Design makes the same point in quieter ways. Walk into a room with a clear axis, balanced proportions, and consistent light. The body slows. Architects have measured this. Studies of hospital design in the 1980s, including work by Roger Ulrich, showed that patients in rooms with regular visual order and views of nature experienced shorter post-operative stays. The behaviour was reduced stress indicators. The metric was recovery time. The implication was care. Pattern supported healing.

But impose the same order without regard for use, and resistance follows. Modernist housing projects often privileged visual regularity over social reality. Pruitt–Igoe returns as warning. Repetition without participation decays. The metric, again, was abandonment. The implication was failure.

The quiet motif threading these examples is the metronome. When it sets a shared tempo, musicians play together. When it dominates, expression flattens. The metronome is useful. It is not sovereign.

Communities negotiate this balance continually, often without naming it. Consider weekly rest. In 321 CE, Emperor Constantine declared Sunday a day of rest in the Roman Empire. The decision formalised an emerging Christian pattern. The behaviour was cessation of labour. The metric was compliance across provinces. The implication was social synchrony. Time itself became patterned.

Later traditions adapted the principle differently. Jewish Sabbath observance, codified in rabbinic texts compiled between the third and sixth centuries CE, specified detailed prohibitions. The structure was intricate. The visible behaviour was abstention from defined activities. The metric was observance fidelity. The implication was identity. The pattern marked belonging.

Yet debates within those communities show the cost of over-specification. Where boundaries multiply, transgression multiplies with them. The reconciliation is pragmatic. Patterns must be strict enough to hold meaning and loose enough to breathe.

Modern life intensifies the stakes because patterns now arrive pre-packaged and personalised. Work schedules fragment. Notifications interrupt. Sleep cycles shift. The metronome stutters. Psychological research, including a 2017 study by Matthew Walker on sleep regularity, links irregular schedules to impaired mood and cognition. The behaviour is variable bedtimes. The metric is sleep efficiency. The implication is fatigue.

Against this backdrop, the appeal of ritual is not nostalgic. It is physiological. Regularity repairs. But the repair only lasts if the pattern is entered freely. Compulsion corrodes its effect.

The discomfort that pattern provokes is therefore not accidental. It signals a question. Who sets the tempo? When repetition aligns with consent, it steadies. When it overrides judgement, it provokes resistance. This is as true of prayer as it is of bureaucracy.

One more counter-example guards against romanticism. In certain high-control religious movements, repeated practices are used to narrow attention deliberately. Former members often report identical behaviours: scripted speech, fixed schedules, restricted movement. The metric is behavioural conformity. The implication is control. Pattern here is not neutral. It is weaponised.

This does not indict repetition. It indicts intent. The difference lies in whether a pattern invites return or forbids departure. The former builds trust. The latter enforces it.

Measure again. Compare again. Commit carefully.

The unresolved tension persists. When you feel comforted by a pattern, do you ask who benefits from its stability? When you feel constrained, do you ask whether flexibility is possible without collapse? The metronome keeps ticking. The decision is how closely to follow it.

Meaning, not mere decoration

In around 800 CE, on the island monastery of Iona off the western coast of Scotland, Columban monks completed what is now known as the *Book of Kells*. The manuscript's pages explode with interlaced knots, spirals, and zoomorphic forms so dense that letters dissolve into pattern. When the book was later moved to Kells in County Meath for safekeeping, it already carried a concrete outcome: it functioned as a Gospel book so revered that it was displayed more than read. The decoration was not marginal. It was the message, rendered visible.

Look closely and the ink refuses to behave. Lines loop back on themselves. Animals bite their own tails. Crosses multiply within crosses. This is not embellishment added after meaning has been settled. It is theology compressed into form. The page teaches before the text is deciphered.

Count the turns.

The assumption that religious pattern exists to please the eye is a modern one, shaped by habits of separation: art here, doctrine there. Historically, the separation rarely held. In many traditions, pattern functioned as a parallel language, capable of carrying ideas that resisted linear statement. The visible behaviour was patient labour. The metric was hours per page. The implication was reverence. Time itself became an offering.

Manuscript illumination provides a clear case. The *Book of Kells* was not unique. From the Lindisfarne Gospels produced in Northumbria around 700 CE to Qur'anic manuscripts copied in Abbasid Baghdad in the ninth century, sacred texts were framed by pattern that instructed the reader how to approach them. In Islamic calligraphy, where figurative imagery was often avoided, geometric borders and vegetal arabesques proliferated. The abstraction was divine unity. The visible behaviour was endless repetition without focal point. The metric was symmetry and continuity. The implication was transcendence. God exceeds depiction; order gestures instead.

The same logic appears in textile traditions. In Andean cultures, long before European contact, woven patterns encoded social identity and cosmology. Archaeological studies of Paracas textiles dating to around 200 BCE show repeated motifs whose arrangement signalled lineage and ritual role. The behaviour was weaving according to inherited schemes. The metric was motif placement. The implication was belonging. To wear the cloth was to inhabit a story.

These patterns were not private expressions. They were legible to those trained to see them. Meaning resided in recurrence and variation, not in novelty. A spiral repeated signalled continuity; a break signalled transition. The cloth taught silently.

Architecture amplifies the lesson at scale. In 1137, work began on the rebuilding of the Abbey Church of Saint-Denis near Paris under the direction of Abbot Suger. Suger's innovations in proportion, light, and verticality are often credited with inaugurating Gothic architecture. His writings make the intention explicit. Measured harmony, he argued, lifted the mind from the material to the immaterial. The abstraction was ascent. The visible behaviour was craned necks and slowed steps. The metric was vertical ratio. The implication was devotion shaped by space.

Here pattern directs the body. Repeating columns establish rhythm. Vaults distribute weight predictably. Light enters through measured openings. The building becomes a diagram one walks through. The theology is not spoken. It is felt.

The motif returns: the curve, the grid, the interval.

To say that pattern teaches does not mean it speaks unambiguously. Learning to decode it requires context. Without that, pattern can be misread or emptied of significance. A limitation emerges. When motifs migrate without their grammar, they become aesthetic residue.

Consider the mandala. In Tibetan Buddhism, mandalas function as ritual diagrams, used in meditation to map the cosmos and the mind. The behaviour is construction and dissolution. Sand mandalas, painstakingly assembled over days, are swept away upon completion. The metric is duration versus erasure time. The implication is impermanence. The pattern teaches by vanishing.

When mandala designs appear on mass-produced wall hangings divorced from practice, the lesson changes. The visible behaviour is display. The metric is permanence. The implication is confusion. Pattern becomes décor when its discipline is removed. The counter-example clarifies the rule. Meaning depends on use, not appearance alone.

This dependence complicates any attempt to universalise symbols. A spiral in a Celtic manuscript and a spiral in a corporate logo share geometry, not theology. The reconciliation lies in attention to function. What does the pattern ask of the viewer? Does it demand patience, repetition, humility? Or does it merely attract?

Return to architecture for a measured comparison. Mosques built under the Seljuk and later Ottoman empires employed repeating geometric tiles with mathematical precision. In the Süleymaniye Mosque in Istanbul, completed in 1557 under the architect Mimar Sinan, the patterning directs attention toward the mihrab without depicting figures. The abstraction is orientation toward Mecca. The visible behaviour is aligned prayer rows. The metric is axial symmetry. The implication is unity of direction.

The pattern does not replace doctrine. It reinforces it through spatial constraint. Bodies align because walls guide them. The decision to kneel is shared. Pattern becomes coordination.

Christian iconography offers a different but related case. Byzantine icons employ strict compositional rules: inverse perspective, gold backgrounds, repeated gestures. The Seventh Ecumenical Council at Nicaea in 787 CE affirmed the veneration of icons, distinguishing honour paid to the image from worship due to God. The visible behaviour was kissing and lighting candles. The metric was regulated depiction. The implication was mediation. Pattern protected theology from idolatry by fixing form.

Here rigidity serves a purpose. Freedom is constrained to preserve meaning. The risk is clear. When form hardens without reflection, icons become clichés. The tradition responded by renewing attention to technique and prayerful preparation. Pattern required discipline to remain alive.

The same dynamic appears in music notation. Gregorian chant, standardised in the early medieval period, relies on repeated melodic patterns known as neumes. The abstraction is sacred time. The visible behaviour is unison singing. The metric is pitch interval repetition. The implication is collective voice. The chant teaches restraint. No soloist dominates.

When later musical styles introduced harmony and rhythm borrowed from secular forms, debates erupted. Was meaning diluted or enriched? The answer varied by community. The pattern argument persisted. What forms can carry weight without collapse?

The quiet motif threads again. The page. The cloth. The wall. Each bears marks that are not random. Each demands decoding. The demand is not intellectual alone. It is ethical. To read pattern well is to accept responsibility for interpretation.

Modern viewers often approach sacred artefacts as museum pieces. The behaviour is observation behind glass. The metric is time spent looking. The implication is distance. Meaning recedes. This is not inevitable. It is a choice enforced by context.

A counter-example emerges in contemporary synagogue architecture that integrates ancient pattern with modern use. The Hurva Synagogue in Jerusalem, rebuilt and reopened in 2010, employs repeated arches and geometric floor designs that echo earlier structures destroyed in 1948. The visible behaviour is gathering under a restored dome. The metric is continuity of form across rupture. The implication is memory encoded in space.

Here pattern functions as condensed history. It does not erase loss. It frames it.

The claim, then, is not that all pattern is sacred, but that religious traditions have consistently treated pattern as a carrier of meaning that words alone cannot sustain. Decoration is surplus. Teaching is embedded. To dismiss the former is to miss the latter.

This reading demands effort. Pattern does not announce its content. It waits. The waiting is part of the lesson. Speed flattens form. Attention restores it.

The unresolved tension remains practical. When encountering a patterned space or object, do you ask what it demands of time, body, and behaviour? Or do you register it as surface and move on? The curve is there, still turning. The question is whether you follow its line or treat it as ornament.

Chapter 2
The Grammar of Shape

In the summer of 1798, Napoleon Bonaparte stood with his engineers on the Giza plateau, gazing at the pyramids. According to later accounts, he gestured toward the stone triangles and remarked that forty centuries looked down upon his army. The line has been rehearsed to the point of exhaustion, yet the moment itself remains instructive. Napoleon was not awed only by age or scale. He was confronted by shape. The pyramids' power lay not merely in their mass, but in their clarity: four lines rising to a point, a geometry so spare that it survives translation across cultures, religions, and empires. The form spoke before history did.

That encounter captures a quiet truth most of us absorb without noticing. Certain shapes do not wait to be explained. They register immediately. A straight line signals direction. A circle encloses. A triangle stabilises or threatens, depending on its stance. We read these forms faster than words. We trust them or recoil from them before reflection intervenes. Shape operates as a language already half-known.

This chapter begins from that recognition and presses it further. If pattern is the universe's handwriting, then shape is its grammar: the minimal set of elements from which more complex meanings are built. Lines, circles, angles, proportions—these are not abstract curiosities but the bones of space. They structure how we build, how we gather, how we move, and how we imagine what cannot be seen directly. Long before doctrines are stated, shapes have already framed belief.

The simplicity is deceptive. A line seems innocent enough, yet it carries moral weight. In maps, lines divide territory and decide allegiance. In architecture, they channel movement, separating public from private, sacred from ordinary. A long, unbroken axis invites procession; a jagged one resists it. The emotional charge is measurable. People walk differently down straight corridors than through curved ones. Bodies sense intention.

Circles do something else entirely. They gather rather than direct. A circle has no obvious beginning or end, which is why it has so often been used to suggest completeness, eternity, or protection. From prehistoric stone rings to baptismal fonts, the form recurs with remarkable persistence. Stand in a circular space and hierarchy softens. No position dominates. Attention turns inward or upward. The shape teaches without speaking.

Triangles introduce tension. They point. They anchor. They imply relation. In flags, triangles often signal movement or aspiration; in religious art, they have been pressed into service as visual shorthand for triads, covenants, or balances of power. A triangle resting on its base feels settled. One inverted feels unstable. The nervous system notices the difference instantly.

These responses are not culturally arbitrary. They are rooted in how human perception evolved to navigate space. Edges matter. Enclosures matter. Balance matters. The grammar of shape is learned by walking, reaching, and avoiding danger long before it is formalised in geometry. When later traditions invest these forms with symbolic meaning, they are not inventing from nothing. They are refining instincts already present.

Symmetry sharpens the point. Faces with bilateral symmetry are consistently rated as more attractive across cultures, a finding replicated in multiple psychological studies. Flowers arrange petals radially because it works: pollinators find them more easily, growth stresses distribute evenly. Temples mirror themselves because symmetry signals care. Something symmetrical announces that effort has been expended, that intention has intervened. The visible order invites trust.

Yet symmetry also raises a philosophical problem. When we find it in nature, are we witnessing design or efficiency? The honeycomb offers a famous case. Hexagonal cells maximise storage while minimising wax, a fact demonstrated mathematically by Thomas Hales in 1999. Bees are not theologians. They follow local rules. The result is elegance without foresight. The question lingers. Does the beauty reside in purpose, or does purpose merely borrow beauty's shape?

This ambiguity is not a flaw. It is the engine of reflection. Religious traditions have long hovered at this boundary, reading natural symmetry as sign without claiming to exhaust its cause. A flower's radial order can be celebrated as creation's generosity without denying biology. The grammar allows multiple readings, provided one resists forcing it into a single explanatory mould.

Proportion deepens the grammar. Humans respond strongly to certain ratios, particularly those that balance repetition and variation. In music, simple intervals soothe; extreme dissonance unsettles. In architecture, rooms that adhere to familiar proportions feel inhabitable; those that violate them feel either exhilarating or oppressive. The nervous system registers these differences before they are articulated. Balance is felt.

This sensitivity explains why proportions recur across time and medium. It is not that a single ratio holds a mystical monopoly on beauty, but that ranges of proportion align with how perception integrates information. Too much sameness dulls attention. Too much variation fragments it. Harmony sits between. Cultures discover this repeatedly, sometimes independently, sometimes through transmission. The result is a family resemblance of forms rather than a universal formula.

When geometry crosses from description into symbol, the stakes rise. A diagram stops being a tool and becomes a teaching. Mandalas map not only space but stages of attention. Star patterns trace relationships that resist linear explanation. The act of drawing, tracing, or walking these forms becomes a discipline. Geometry here is not an illustration of belief; it is a method of forming it.

This chapter follows that crossing carefully. It asks when and why communities agree that a shape points beyond itself. Agreement matters. A symbol is not private. It functions because it is shared, because its grammar is taught and rehearsed. Without that shared literacy, forms revert to ornament. With it, they become condensed arguments about reality.

Modern life complicates this inheritance. Shapes circulate freely now, detached from the contexts that once trained their interpretation. Sacred diagrams appear on yoga mats, corporate logos, and digital wallpapers. Some of this circulation is creative, some careless. The grammar persists, but fluency declines. People feel the pull of form without knowing why. The result is both opportunity and risk.

The opportunity lies in recovery. Learning the grammar of shape does not require nostalgia or dogma. It requires attention. When one notices how a floor plan organises movement, how a logo signals allegiance, how a screen layout directs choice, the world becomes legible in new ways. Shapes cease to be neutral. They reveal intent.

The risk lies in manipulation. Geometry can persuade without argument. Symmetry can mask injustice. Proportion can naturalise exclusion. A perfectly balanced space may still be closed to some bodies. The grammar of shape, like any language, can be used to tell the truth or to conceal it. Literacy therefore carries responsibility.

This chapter does not promise mastery. It offers orientation. By the end, the reader will have a clearer sense of why simple forms carry such disproportionate weight, and how to ask better questions of the spaces and symbols that surround them. Why does this place feel trustworthy? Why does that image command loyalty? Why does another provoke unease?

The bones of space are always present. We walk among them, sit within them, and project meaning through them whether we intend to or not. To learn their grammar is not to strip the world of mystery. It is to recognise that mystery often arrives already shaped.

The pages that follow move from line to circle to proportion, not as abstract lessons but as encounters. They invite the reader to see how much has already been said without words, and how much remains to be heard once the language of shape is taken seriously.

Lines, circles, and the bones of space

In December 1946, at Lake Success in New York, the newly formed United Nations adopted its emblem. The design decision was deliberate and recorded. A world map, flattened and enclosed by olive branches, would represent the organisation across documents, buildings, and peacekeeping missions. The outcome was immediate and practical: a symbol legible across languages, reproduced at scale, and trusted enough to mark vehicles and personnel in conflict zones. The form did the work. The emblem spoke before any charter was read.

It is only a circle. It is only lines. Yet the choice mattered.

Shapes persuade quickly.

This is the quiet fact from which this chapter proceeds. Simple geometric forms act as a basic alphabet long before they are assembled into complex arguments. A line draws attention and directs it. A circle gathers and contains. A triangle balances forces or points toward risk. These forms carry emotional and symbolic weight not because cultures agree to assign meaning arbitrarily, but because bodies respond to them reliably. Icons, flags, and floor plans exploit this reliability with care or with consequence.

Begin with the line. Stand in a long corridor and your pace changes. Shoulders square. Eyes orient forward. Architects have measured this. Straight axes increase walking speed and reduce hesitation at decision points, a finding reported in wayfinding studies across transport hubs in the 1990s. The behaviour is directional movement. The metric is time-to-destination. The implication is authority. Lines command.

Religious architecture has used this command for centuries. Processional routes in cathedrals draw worshippers toward an altar placed at the terminus of a clear axis. The line becomes pilgrimage compressed. In political contexts, the same form asserts control. Parade grounds extend lines to choreograph bodies. The shape is neutral. Its use is not.

The circle operates differently. It resists direction. Gather people in a circle and hierarchy softens. Eye contact distributes evenly. Anthropologists have noted this repeatedly in council practices, from Indigenous governance in North America to monastic chapter meetings in Europe. The behaviour is turn-taking. The metric is speaking-time variance. The implication is inclusion.

The UN emblem relies on this effect. By enclosing the map in a circle, it removes the sense of front and back. No nation sits above another. The olive branches repeat the enclosure, reinforcing protection. The visible behaviour follows. Flags bearing the emblem are treated as neutral ground. The metric is incident rate around marked sites. The implication is trust, fragile but real.

Triangles complicate matters. They introduce tension. Three points establish relation and stability simultaneously. Engineers rely on triangles because they do not collapse under load. The behaviour is load distribution. The metric is structural integrity. The implication is reliability.

Symbolically, triangles have been pressed into service wherever balance among forces matters. In Christian iconography, triangular compositions stabilise scenes of tension. In national flags, triangles often mark aspiration or movement. The flag of Nepal, adopted in its current form in 1962, remains the only non-rectangular national flag. Its stacked triangles reference the Himalayas and the two major religions of the country. The visible behaviour is vertical orientation. The metric is recognisability. The implication is identity anchored in terrain.

These forms recur because they work. The nervous system registers them quickly. Cognitive psychologists studying visual perception have shown that straight edges, closed curves, and pointed angles are detected faster than irregular shapes. The behaviour is saccadic eye movement. The metric is detection latency. The implication is efficiency. Shape saves time.

Icons exploit this economy. A halo is a circle for a reason. It separates the figure from background without enclosing the body. The circle floats. The eye understands sanctity before narrative. Gold backgrounds flatten depth, replacing spatial realism with symbolic clarity. The decision to use these forms was not aesthetic whim. It was pedagogical.

Floor plans teach with the same economy. A basilica's nave directs movement forward. A centrally planned church gathers it inward. The difference is measurable. Studies of congregational movement show longer dwell times in centralised spaces and faster processional flow in axial ones. The abstraction is reverence. The visible behaviour is pacing. The implication is attention shaped by geometry.

Flags distil these lessons further. Consider the Swiss flag, formalised in 1889: a white cross on a red square. The cross is orthogonal, composed of intersecting lines that meet at a centre. The square stabilises it. The behaviour is instant recognition. The metric is recall accuracy in cross-cultural studies. The implication is neutrality with firmness. The form communicates restraint without passivity.

A counter-example sharpens the point. When symbols ignore bodily response, they falter. In the 1970s, several corporate logos adopted overly complex forms, layering curves and gradients that reproduced poorly and confused viewers. Redesigns followed. The behaviour was misrecognition. The metric was brand recall. The implication was failure of the alphabet. Too many letters obscure the word.

This limitation does not reduce geometry to branding. It clarifies the conditions under which shape carries meaning. Simplicity matters, but so does fit. A circle that encloses too tightly suffocates. A line that extends without pause exhausts. A triangle inverted unnerves. These effects are not mystical. They are somatic.

Religious traditions have navigated this somatic grammar with care. In Hindu temple architecture, the *vastu purusha mandala* lays out space using a grid that aligns cardinal directions with divine attributes. The behaviour is circumambulation along prescribed paths. The metric is ritual sequence. The implication is cosmic order mapped onto ground.

In Islamic prayer halls, the absence of pews preserves openness, while the mihrab—a niche indicating direction—

anchors attention. The line toward Mecca is marked, not enforced. Bodies align voluntarily. The metric is orientation consistency. The implication is unity through choice.

The quiet motif returns: the threshold. Lines mark thresholds. Circles guard them. Triangles announce them. A doorway framed by straight edges signals entry. A circular well invites pause. A triangular pediment warns of weight above. The threshold is where geometry becomes ethical, asking whether to cross.

Modern environments multiply thresholds without clarifying them. Open-plan offices erase lines in the name of flexibility, then reintroduce them digitally through schedules and permissions. The behaviour is constant recalibration. The metric is interruption frequency. The implication is fatigue. Geometry removed from walls returns through screens.

This return underscores the claim. Shapes are not optional. When ignored, they reappear elsewhere. The task is not to eliminate geometry but to read it honestly.

Consider a hospital floor plan. Clear sightlines reduce anxiety. Rounded corners slow movement, reducing collision risk. Designers measure this. The behaviour is walking speed. The metric is accident rate. The implication is care expressed through form. When cost-cutting replaces curves with sharp turns, outcomes shift. Geometry carries consequence.

Art offers another check. Abstract painters in the early twentieth century stripped representation down to line and shape to test their expressive limits. Piet Mondrian's grids, developed in Paris and later New York between 1917 and 1944, rely on vertical and horizontal lines balanced by blocks of primary colour. The behaviour is sustained viewing. The metric is compositional stability. The implication is order without narrative. The painting holds because the grammar holds.

Yet abstraction also reveals a boundary. Without shared context, forms can lose purchase. A grid may signify order to one viewer and sterility to another. The reconciliation lies in use. Geometry teaches best when embedded in practice, not isolated on a wall.

The bones of space are therefore not merely visual. They are behavioural. Lines ask us to move. Circles ask us to stay. Triangles ask us to attend. These requests are felt, not argued.

The unresolved tension is practical. When you enter a space or encounter a symbol, do you ask what its shapes are asking you to do? To proceed, to gather, to brace? Or do you accept the request without noticing it? The alphabet is already in place. The question is whether you read it, or let it read you.

Symmetry as a sign of intention

In 1648, in Agra, the Mughal emperor Shah Jahan completed the Taj Mahal. Contemporary court chronicles record the outcome with unusual clarity: the mausoleum's perfect bilateral symmetry, mirrored across a central axis, was intended to honour Mumtaz Mahal with a form that refused imbalance. Every minaret, garden path, and reflecting pool was aligned so that the structure would read as one composed act. Visitors noticed. They still do. The building arrests movement. People slow, centre themselves, and look twice.

Symmetry commands attention.

That command is not subtle. It operates before interpretation. Stand before the Taj Mahal or any structure that mirrors itself cleanly and the body registers deliberation. Nothing appears accidental. The mind infers care. The inference is fast and stubborn.

Symmetry functions as a signal of intention because it is costly. To make something symmetrical requires constraint, measurement, and repeated correction. Errors show immediately. In a world where irregularity is cheap, symmetry advertises effort. This is why it has been trusted across domains as a marker of seriousness, from human faces to flowers to temples. The trust is not sentimental. It is behavioural.

Begin with faces. In 1999, a study by David Perrett and colleagues at the University of St Andrews demonstrated that faces with higher degrees of bilateral symmetry were consistently rated as more attractive across test groups. The behaviour was selection preference. The metric was rating score variance. The implication was evolutionary: symmetry correlates with developmental stability. Faces that mirror themselves signal resistance to environmental stressors during growth.

The signal is imperfect. No face is truly symmetrical. Close inspection reveals small asymmetries everywhere. Yet the preference holds within limits. Extreme symmetry, digitally exaggerated, produces uncanny discomfort. The counter-example matters. Perfect symmetry, divorced from organic variation, alarms rather than reassures. The reconciliation is precise. Symmetry must coexist with life's noise to persuade.

Flowers offer a cleaner case. Radial symmetry dominates because it works. A daisy presents its petals evenly to pollinators approaching from any direction. The behaviour is pollinator landing frequency. The metric is successful pollination rate. The implication is efficiency. Plants that distribute access evenly gain advantage. Here, symmetry is not aesthetic choice but survival strategy.

Humans notice anyway. Floral symmetry has shaped decorative arts for millennia because it reads as generosity. Nothing is hidden. Access is equal. The eye moves easily around the centre. The circle returns.

Temples scale the principle up. Bilateral symmetry in sacred architecture stabilises movement and expectation. Approach a symmetrical façade and the path becomes legible. Entrances announce themselves. Axes align. In ancient Greek temples such as the Parthenon, completed in 432 BCE in Athens, symmetry was refined rather than absolute. Subtle curvatures corrected optical illusion, preserving the appearance of balance. The behaviour was slowed procession. The metric was alignment accuracy. The implication was reverence engineered through precision.

Religious builders learned quickly that symmetry does more than please. It disciplines. When a space mirrors itself, bodies mirror one another. Rows form. Sightlines converge. Collective attention synchronises. The abstraction is unity. The visible behaviour is coordinated posture. The metric is variance reduction in movement. The implication is order felt as calm.

Yet symmetry also raises a philosophical provocation. When we encounter it in nature, what are we seeing? Is symmetry evidence of design, or the by-product of efficient laws? Snowflakes offer the classic puzzle. In 1611, Johannes Kepler published *Strena seu de Nive Sexangula,* observing that snow crystals consistently form hexagonal shapes. He proposed packing efficiency as explanation. Later crystallography confirmed the role of molecular bonding angles. The behaviour is sixfold repetition. The metric is lattice structure. The implication is necessity, not intention.

And still, people marvel. The reaction persists even when the mechanism is known. This persistence reveals something important. Symmetry does not lose its expressive power when explained. Knowledge reframes wonder; it does not dissolve it. The sign remains legible.

Markets provide an instructive contrast. Price charts rarely display symmetry. Peaks and troughs skew. Recoveries lag. When symmetry does appear—such as mirrored boom-and-bust cycles—commentators infer manipulation or artificial constraint. The behaviour is suspicion. The metric is anomaly detection. The implication is distrust. In systems assumed to be organic, symmetry reads as imposed.

This reversal clarifies the rule. Symmetry signals intention only where intention is plausible. In architecture and ritual, intention is expected. In weather and finance, it raises questions. Context governs interpretation.

The quiet motif threading these examples is the mirror. A mirror duplicates without understanding. Symmetry is a controlled mirror, introduced deliberately or enforced by law. Where mirrors belong, symmetry reassures. Where they intrude, it unsettles.

Temple layouts illustrate the distinction vividly. Hindu temples often combine bilateral symmetry with radial expansion, aligning axes with cardinal directions. The behaviour is circumambulation around a centre. The metric is path repetition. The implication is cosmic order enacted bodily. The symmetry is not total; subsidiary shrines introduce variation. The pattern holds because it breathes.

Islamic architecture often privileges symmetry without central figuration. The Great Mosque of Córdoba, expanded between the eighth and tenth centuries in Spain, employs repeating arches that mirror across multiple axes. The behaviour is steady forward movement interrupted by visual rhythm. The metric is column spacing. The implication is infinity suggested through repetition. Intention is present, but it points beyond the builder.

Christian basilicas, by contrast, often enforce a dominant axis from entrance to altar. Symmetry frames the journey. The behaviour is linear procession. The metric is axis length. The implication is pilgrimage compressed into minutes.

These differences matter. They show that symmetry is not a single message. It is a grammatical tool that can organise attention toward centre, path, or field. To read it well is to ask not merely whether symmetry exists, but how it operates.

A limitation presses here. Excessive symmetry can erase distinction. In bureaucratic environments, symmetrical layouts may communicate interchangeability rather than care. Identical cubicles mirror one another endlessly. The behaviour is disengagement. The metric is dwell time at desks. The implication is anonymity. Symmetry without narrative flattens experience.

Designers respond by breaking symmetry strategically. A single irregular element restores orientation. The reconciliation mirrors the earlier lesson from faces. Life requires asymmetry to remain legible.

The same holds in worship. Liturgies repeat, but not perfectly. Feasts interrupt fasts. Silence interrupts speech.

The mirror is cracked on purpose. Absolute symmetry would suspend time. Human practice requires sequence.

Modern science complicates the design question further. Fundamental physics reveals deep symmetries—conservation laws tied to invariance under transformation. Emmy Noether's theorem, published in 1918 in Göttingen, demonstrated that symmetry principles correspond to conserved quantities. The behaviour is mathematical invariance. The metric is equation stability. The implication is profound. Symmetry underwrites lawfulness itself.

Does this settle the matter? Hardly. Laws describe. They do not intend. Yet the language of symmetry persists because it names reliability. When something remains unchanged under reflection or rotation, it can be trusted to endure. Trust here is not belief. It is expectation borne out by repetition.

The ethical edge returns. When leaders build symmetrical monuments, they borrow that expectation. Power cloaks itself in balance. The risk is manipulation. History offers warnings. Totalitarian regimes have favoured monumental symmetry to project inevitability. The behaviour is mass alignment. The metric is scale. The implication is dominance disguised as order.

Discernment therefore becomes essential. Symmetry deserves respect, not surrender. It invites attention but demands scrutiny. Who paid its cost? Who benefits from its calm?

The unresolved tension remains personal. When you encounter symmetry—on a face, in a flower, within a hall—do you register intention and stop there? Or do you ask what kind of intention is being signalled, and to what end? The mirror stands ready. The decision is how long to look.

Ratio, proportion, and the taste for harmony

In 1863, in Paris, Hermann von Helmholtz published *Die Lehre von den Tonempfindungen als physiologische Grundlage für die Theorie der Musik* (*On the Sensations of Tone*). It was not a composer's manifesto but a physicist's intervention. Helmholtz measured how vibrating strings produce overtones and why certain intervals sound consonant while others grate. The outcome reached beyond laboratories: it gave musicians and instrument-makers a physiological account of harmony, linking ratios of frequency to felt experience. Music, long treated as a mysterious gift, acquired a measurable backbone.

Ratios move the body.

That sentence can be tested. Play two notes whose frequencies stand in a simple relationship—2:1 for the octave, 3:2 for the perfect fifth—and many listeners report ease, even without musical training. The behaviour is stable attention and lowered aversion. The metric, in experimental settings, is reduced reported roughness and faster preference decisions. Helmholtz's contribution was to show why: simple ratios align partials, reducing acoustic beating that the ear interprets as harshness. Harmony is not a metaphor. It is mechanics felt.

The same logic extends beyond sound. Proportion is the arithmetic of fitting. A door that is too narrow is not merely inelegant; it alters gait. A room with awkward dimensions does not merely offend taste; it changes where people stand and how long they stay. The nervous system makes judgements about proportion quickly because proportion predicts action. Can I reach? Can I turn? Can I see? Balance is practical before it is aesthetic.

Show first. Walk into a well-proportioned space and conversation spreads. People cluster without crowding. In a badly proportioned one, bodies press against walls or block thresholds. Name it: affordance, the term used by the psychologist James J. Gibson in 1979 to describe action possibilities offered by an environment. Explain it: proportions cue the body about what is possible. Implication: what we call "taste" often begins as embodied prediction, refined into culture.

Art has always exploited this. In the fifteenth century, Leon Battista Alberti wrote *De re aedificatoria,* published posthumously in 1485, arguing that beauty arises from the harmony of parts to whole. Alberti was not quoting spirits; he was borrowing a musical insight and transferring it to buildings. The abstraction was beauty. The visible behaviour was lingering and return. The metric, for patrons, was reuse: designs that were copied and commissioned again. The implication was economic as well as aesthetic. Proportion earns trust.

Trust shows itself. Clients pay. Communities adopt. Builders repeat.

Yet proportion is not a single recipe. Here, myth needs trimming. The so-called "golden ratio" is often invoked as a universal key to beauty, attributed to the Parthenon, Renaissance painting, even the human face. The claims tend to outrun evidence. Historians of mathematics have noted that many attributions were made after the fact, often by enthusiasts searching for one number to explain diverse successes. This is the limitation that protects the argument from becoming a numerological parlour game. Ratios recur, but not as a monopoly.

The reconciliation begins with a more honest observation. Humans respond to ranges of proportion that balance predictability with interest. Too simple and the mind grows bored. Too complex and the mind grows strained. Between those extremes lies a zone in which attention stays engaged without fatigue. This is not mystical; it is cognitive economy.

In music, this economy is measurable. The octave is universally salient because it doubles frequency and aligns harmonic series. The perfect fifth, 3:2, is likewise stable because it produces minimal beating. But not all cultures privilege the same scales. Ethnomusicologists have documented systems that use microtones unfamiliar to

Western ears. The behaviour is training-dependent preference. The metric is listening familiarity. The implication is cultural plasticity built on biological constraints. The ear provides a set of affordances; cultures choose within them.

Design operates similarly. Proportions that "feel right" often sit near simple fractions: halves, thirds, quarters. This is why grids dominate graphic design and architecture. A half is easy to anticipate. A third offers variation without chaos. The body and eye can track the relationships without strain. The abstraction is clarity. The visible behaviour is faster navigation. The metric is time-to-find in usability testing. The implication is humane efficiency.

The modern world provides a vivid laboratory. In 1922, the architect Le Corbusier proposed regulating proportions in architecture through a system he later formalised as the Modulor (published in 1948 in France). He aimed to anchor design in the human body, using measured relationships between height, reach, and spatial intervals. The outcome was concrete: the Modulor influenced buildings, furniture, and industrial design, shaping dimensions that still feel "standard." The behaviour is comfortable reach and circulation. The metric is clearance and ergonomic fit. The implication is that proportion can be institutionalised.

Institutionalised proportion comforts precisely because it reduces uncertainty. Standard sizes mean fewer surprises. Doors fit bodies. Steps fit legs. Chairs fit hips. The nervous system relaxes when it can predict. Yet the same standardisation can also flatten. When every room conforms, difference disappears. The limitation returns. Proportion can become a cage.

Consider the urban grid. In Manhattan, the Commissioners' Plan of 1811 imposed a rectilinear street grid across the island. The ratio of block lengths and widths made navigation easy and development profitable. The behaviour is direct movement. The metric is route predictability. The implication is power. A grid makes land legible for commerce.

But legibility is not the same as liveability. Critics from Jane Jacobs onwards have shown how uniform grids can ignore local social ecologies. The behaviour is displacement. The metric is rising rents and demographic shift. The implication is moral: proportions that serve capital may not serve community. The taste for harmony must therefore be tested against consequences.

Return to music for a parallel. Equal temperament, widely adopted in Europe by the nineteenth century, divides the octave into twelve equal semitones. It allows instruments to play in any key without retuning. The outcome was enormous: it enabled the harmonic complexity of later Western music and standardised instrument manufacture. The abstraction is flexibility. The visible behaviour is modulation between keys. The metric is tuning compromise. The implication is trade-off. Equal temperament sacrifices perfect ratios in any one key for functional adaptability across all.

This is proportion's deeper lesson. Harmony is not purity; it is negotiated fit. The ear tolerates slight impurity for the sake of mobility. The same negotiation appears in buildings. Perfect symmetry may be sacrificed for circulation. Ideal ratios may be compromised to admit light. The "taste" for harmony is often a taste for workable compromise that still feels coherent.

Neuroscience offers evidence that coherence is not an abstract preference but a regulation strategy. Research on predictive processing—popularised in neuroscience by Karl Friston from the mid-2000s onwards—frames the brain as a system that minimises prediction error. Proportions that align with bodily expectation reduce error. The visible behaviour is reduced startle and hesitation. The metric is smoother movement and lower reported discomfort. The implication is that what we call balance is partly the nervous system conserving effort.

This conservation does not erase the spiritual dimension. In sacred architecture and ritual, proportion often becomes a deliberate moral training. A measured space asks the body to slow. A repeated interval in chant asks the mind to stay. The ratio becomes a discipline. In this sense, proportion is not simply pleasing; it is formative.

Yet spiritual traditions also warn against mistaking the instrument for the aim. If proportion becomes fetish, it displaces meaning. A perfectly tuned chant sung without attention is empty. A flawlessly proportioned shrine used to exclude the vulnerable is a lie dressed as harmony. The counter-example is ethical. Proportion can be used to launder power.

So the question returns in a sharper form. Why do certain ratios recur in art, music, and design? Because they are easy for bodies and minds to inhabit. They reduce friction. They align expectation with experience. But this very ease makes them persuasive, and persuasion needs scrutiny.

The quiet motif threading this discussion is the interval: the space between notes, between walls, between actions. Intervals determine whether a melody resolves, whether a corridor invites, whether a life breathes. When an interval is too tight, pressure rises. When it is too loose, coherence dissolves. The right interval does not announce itself. It simply holds.

Hold is the word. A good proportion holds attention without gripping it. It holds a community without imprisoning it. It holds a thought long enough for understanding to form.

The forward test is not mathematical. It is practical. When a space, a piece of music, or an image feels "right," do you ask what trade-off made that ease possible—whose needs were centred, whose were compressed, which freedoms were exchanged for fit? The interval remains. The decision is whether to accept it as given or to measure it again.

When geometry becomes symbol

In 779 CE, at Samye Monastery in central Tibet, the Indian master Padmasambhava presided over rituals that included the construction of mandalas—carefully measured diagrams laid out on cloth and ground to guide meditation and teaching. Contemporary Tibetan chronicles record the outcome with precision: monks memorised cosmology by walking it. The diagram was not illustrative. It functioned. The mandala organised attention, posture, and sequence, shaping how doctrine entered the body. Geometry crossed a threshold. It became instruction.

The lines were exact. The silence mattered.

At that point, shape ceased to be a neutral description of space and became a vehicle for meaning. This crossing is not rare, but it is decisive. Geometry becomes symbol when a community agrees that form does not merely depict order but *enacts* it. The agreement is practical, not abstract. People move, pause, and repeat according to the diagram's demands. Behaviour changes. Memory follows.

Show first. A sand mandala takes days to construct and minutes to dismantle. Name it: ritualised impermanence. Explain it: the act trains attention to hold complexity lightly. Implication: knowledge is not possession but practice. The geometry teaches by being used, not by being admired.

This transition—from measurement to meaning—has occurred repeatedly across cultures. Diagrams that begin as explanatory tools harden into emblems when they are trusted to carry truth across generations. The trust is visible. People return to the same shapes. They copy them faithfully. They correct deviations. Fidelity becomes a moral act.

Consider the Christian cross. As geometry, it is a simple intersection of lines. As symbol, it condenses narrative, ethics, and identity. By the fourth century CE, following Constantine's adoption of the cross after 312 CE in Rome, the form migrated from execution device to emblem. The behaviour shifted. Crosses were traced over bodies. Churches aligned themselves to it. The metric was ubiquity across space. The implication was allegiance. A shape reorganised belonging.

The transformation did not erase geometry; it intensified it. The proportions of the cross varied—Latin, Greek, Celtic—each variation teaching something slightly different about emphasis and reach. The vertical lengthened to suggest transcendence. The arms widened to suggest embrace. Measurement became catechesis.

Islamic star patterns offer a parallel without figuration. In fourteenth-century Granada, artisans of the Nasrid dynasty covered the Alhambra with complex interlacing stars, some employing quasi-periodic tilings that modern mathematicians only fully described centuries later. The behaviour was repeated carving to exact rule. The metric was rotational symmetry count. The implication was inexhaustibility. The pattern could extend indefinitely without repetition, gesturing toward divine infinity.

Here geometry becomes symbol through refusal. No figures appear. No narrative scenes distract. The star pattern teaches by denying focal points. Attention circulates. Contemplation replaces consumption. The visitor does not *read* the wall; the wall reads the visitor.

Diagrams sharpen this dynamic further. Medieval Christian thinkers used geometric figures to teach doctrine where language faltered. The Shield of the Trinity, attested in twelfth-century manuscripts in France and England, arranged the relations of Father, Son, and Spirit in a triangular diagram connected by labelled lines. The abstraction was relational ontology. The visible behaviour was tracing lines while reciting propositions. The metric was memorisation accuracy. The implication was orthodoxy stabilised through spatial logic.

The diagram's power lay in constraint. It limited what could be said while preserving complexity. Geometry here functioned as a governor on speculation. It allowed thought to move without drifting. This is symbol at work as a brake as much as an accelerator.

The quiet motif threading these cases is the path. Mandalas are walked. Crosses are traced. Stars lead the eye along set routes. A symbol differs from decoration because it prescribes movement. The body participates.

This prescription explains why geometry becomes sacred only under certain conditions. A beautiful pattern alone does not suffice. The community must *use* it. Use leaves marks. Paths wear into floors. Fingers polish stone. The metric is erosion. The implication is longevity earned through repetition.

A credible limitation presses here. Symbols can ossify. When the use becomes rote, geometry loses teaching force and becomes badge. The behaviour is mechanical repetition. The metric is disengagement, measured in shortened attention or declining participation. The implication is hollowing. A star carved endlessly may cease to invite thought.

History records such failures. In late medieval Europe, complex church floor labyrinths—such as the famous example at Chartres Cathedral, installed around 1200—initially served as surrogate pilgrimages, walked prayerfully when travel was impossible. Over time, many were covered by pews or treated as curiosities. The behaviour shifted from walking to looking. The metric was bodily engagement. The implication was loss of function.

The reconciliation lies not in abandoning symbol but in renewing practice. Where labyrinths are again walked, as they have been since the late twentieth century in churches and hospitals, the teaching returns. Geometry regains voice through feet.

Star patterns illustrate another boundary. In modern design, Islamic geometric motifs are often lifted wholesale and applied to wallpapers or logos. The behaviour is visual consumption. The metric is trend adoption rate. The implication is dilution. Without the disciplines of repetition, orientation, and scale that once governed them, the stars become surfaces.

This is not theft so much as truncation. Meaning drains when use changes. The same shape, under different conditions, speaks differently.

Mandalas demonstrate the point with unusual clarity. In Tibetan practice, their construction follows strict rules transmitted through lineage. The behaviour is apprenticeship. The metric is error correction frequency. The implication is authority grounded in transmission. When mandalas circulate as printed posters detached from practice, the authority evaporates. The form remains. The teaching does not.

The same pattern appears in Hindu yantras—geometric diagrams used in meditation and ritual. The Śrī Yantra, documented in tantric texts by the first millennium CE in South Asia, arranges interlocking triangles around a central point. Practitioners focus attention along prescribed routes. The behaviour is sustained gaze. The metric is concentration duration. The implication is interior alignment. Geometry trains the mind by giving it nowhere else to go.

Scientific diagrams, though secular, sometimes cross into symbolic territory as well. The Bohr model of the atom, proposed in 1913 in Copenhagen, persists in education long after its limitations were known. Electrons do not orbit nuclei like planets. Yet the diagram endures because it teaches relational structure efficiently. The behaviour is conceptual grasp. The metric is explanatory speed. The implication is pedagogy over precision.

Here the counter-example cautions. When diagrams harden into symbols, they risk misleading. The reconciliation requires explicit humility. Symbols must be held as tools, not truths. Religious traditions that remember this tend to refresh symbols through ritual dismantling or periodic reinterpretation.

Why, then, do shapes and ratios become sacred emblems at all? Because they offer a way to carry complexity without exhaustion. A diagram compresses doctrine. A star pattern holds infinity without collapse. A mandala maps stages of attention without text. The nervous system appreciates the compression. The behaviour is return. The metric is endurance across centuries. The implication is that form can outlast explanation.

This endurance creates responsibility. Symbols persuade silently. They bypass argument. A cross on a hill, a star on a wall, a diagram on a page—each frames what can be thought and felt nearby. To deploy geometry as symbol is to shape conscience.

Modern environments deploy symbols constantly, often without acknowledging their power. Corporate logos borrow sacred proportions. Urban plazas adopt mandala-like layouts. The behaviour is foot traffic routing. The metric is dwell time. The implication is influence without consent. Geometry teaches whether or not it is named as teacher.

The unresolved tension therefore sharpens. When you encounter a powerful form—a diagram that steadies you, a pattern that draws you inward—do you ask what practice sustains it and what it asks of you in return? Geometry has crossed the threshold. The question is whether you cross with it, or let it instruct you at a distance.

Chapter 3
Fractals and Infinite Echoes

Stand at the edge of a coastline and look down. The shore twists and breaks into inlets, headlands, and coves. From a distance it appears jagged but coherent, a single contour holding its ground against the sea. Step closer and the same irregularities reappear: smaller bays within bays, miniature peninsulas echoing larger ones. Bend down again and the pattern refuses to resolve. The edge keeps repeating itself. No matter how closely you look, the coastline never quite smooths out.

This is the quiet shock that unsettles first-time observers. Complexity does not dissolve under scrutiny; it reproduces. The detail mirrors the whole. What seemed like an exception reveals itself as a rule.

That rule has a name—fractals—but the name matters less than the experience. Fractals announce a different kind of order, one that does not rely on uniformity or symmetry, but on repetition across scale. They invite a shift in how we imagine structure, growth, and meaning itself. Instead of asking where a form begins and ends, they ask how it keeps going.

This chapter explores that question from several angles. It begins with the visible world, where self similarity appears so frequently that it is easy to overlook. Leaves branch like trees; trees branch like rivers; rivers branch like blood vessels. The forms are not identical, yet they rhyme. A single rule, iterated patiently, generates astonishing variety. Complexity does not require complexity at the outset. It accumulates.

The intuition matters. Fractals are often introduced through equations and computer graphics, but their deeper power lies in what they allow us to see without mathematics. A fern frond repeats the shape of the whole plant at smaller scales. Break a cauliflower and the pieces resemble the original. The eye grasps this immediately. The mind follows more slowly. What looks elaborate is the product of simple constraints applied again and again.

This recognition unsettles older habits of thought. We are accustomed to associating order with regularity and chaos with irregularity. Fractals refuse the distinction. They are ordered and irregular at the same time. Their boundaries are precise but never smooth. They are disciplined without being tidy. In that refusal lies their relevance.

Living systems make the point unavoidable. Lungs branch because branching maximises surface area within limited volume. Roots branch because they must explore soil efficiently without collapsing the plant. River basins branch because water follows gradients and conserves energy. Scaling laws govern these forms, linking size to function across orders of magnitude. A mouse and an elephant do not simply differ by scale; their internal architectures adapt to the same constraints in proportionate ways.

This is not aesthetic coincidence. It is economy written in matter. Energy, time, and material are conserved through repetition. The same solutions reappear because they work. When we notice this, admiration deepens into respect. The universe is not extravagant. It is thrifty.

Yet the fascination with fractals extends beyond biology and geology. Human beings respond to repeating forms with an attentiveness that is both calm and alert. Studies of visual perception have shown that certain fractal patterns reduce physiological stress while holding attention longer than uniform designs. The effect is subtle but measurable. Heart rates stabilise. Eyes linger. The mind neither tires nor rushes ahead.

This response is not accidental. Fractals sit between predictability and surprise. They offer enough repetition to reassure and enough variation to engage. The nervous system recognises the balance. It is the same balance that sustains conversation, music, and ritual. Too much sameness deadens. Too much novelty overwhelms. Fractals inhabit the middle ground.

Artists and builders have exploited this intuition long before it was named. Textiles across cultures employ nested motifs that repeat at different scales. Islamic tilework, Gothic tracery, African weaving patterns—all rely on self similarity to generate richness without chaos. Murals and mosaics draw the eye inward by repeating forms that never quite close. The viewer is invited to stay.

Digital art has rediscovered the appeal with new tools. Computer-generated fractals fill screensavers and visualisations not because they are fashionable, but because they hold attention without exhausting it. They promise depth without demand. In an age saturated with sharp edges and abrupt transitions, this matters.

The chapter does not stop at perception. It presses inward. Fractals offer more than a description of external form; they provide a metaphor for interior life that is difficult to ignore once seen. Human character does not develop in a straight line. Patterns repeat across decades. Old habits resurface in new contexts. Lessons learned once return under different names. Growth looks less like ascent and more like spiralling.

This is not a counsel of despair. Self similarity does not mean stagnation. Each turn occurs at a new scale. What was once overwhelming becomes manageable; what was once invisible becomes obvious. The same challenges reappear, but we meet them differently. Fractals help us see this not as failure, but as structure.

Spiritual traditions have long intuited this rhythm. Pilgrimages circle familiar themes. Prayers repeat with slight inflection. Scriptures return to the same questions—justice, mercy, fidelity—without exhausting them. The repetition is not redundancy. It is depth revealed gradually. The pattern teaches patience.

This chapter holds these threads together without forcing them into a single claim. Fractals are not proofs of doctrine. They are invitations to reconsider how meaning persists without closure. They show how order can be open-ended, how complexity can arise without design in the narrow sense, and how repetition can coexist with change.

Modern relevance presses quietly here. We live amid systems whose complexity feels unmanageable: economies, ecologies, information networks. The temptation is to search for total control or to surrender to confusion. Fractals offer a third posture. They suggest that understanding may come from recognising repeated local rules rather than mastering the whole. Small changes, applied consistently, scale.

The metaphor is demanding. It asks for humility. One cannot grasp a fractal in a single glance. One must zoom, and then zoom again. Understanding is provisional. Each level reveals another. This is not a failure of comprehension but its proper shape.

As the chapter unfolds, it will move from leaf to lung, from textile to mind, from coastline to conscience. It will show how the same logic governs growth in matter and maturity in persons. It will examine why repeating forms calm us and why they sometimes unsettle, especially when they expose patterns we would rather deny.

The aim is not to aestheticise complexity or to romanticise recurrence. It is to offer a lens through which persistence and change can be held together without simplification. Fractals do not promise resolution. They promise continuity.

By the end, the reader will not be asked to memorise definitions or equations. They will be asked to notice. To look at a familiar object and see it echo itself. To recognise in their own life the return of themes, not as traps but as invitations to deepen response. To accept that meaning may not arrive once and for all, but again and again, each time closer, each time slightly altered.

The coastline remains. The closer you look, the more it speaks.

Self similarity in leaf and coastline

In 1967, Benoît Mandelbrot published a paper in *Science* titled "How Long Is the Coast of Britain?" The question sounded whimsical. The outcome was not. Measuring the British coastline, Mandelbrot showed, produces wildly different answers depending on the length of the ruler used. Shorten the measuring stick and the coastline grows longer. Use a finer scale and more inlets appear. The edge refuses to settle. The place was Britain; the date was precise; the implication was unsettling. Natural forms do not converge on a single length because their detail repeats at every scale.

Look closer. It keeps going.

This was not a quirk of cartography. Mandelbrot had exposed a structural feature of reality. The coastline was not messy by accident. It was organised by a rule that generated similar irregularities no matter how far one zoomed in. The name he later gave to such forms—fractals—was less important than the visual intuition they demanded. Complexity, it turned out, could be lawful without being smooth.

The same intuition can be grasped without mathematics. Pick up a leaf and examine it. The central vein branches into smaller veins, which branch again into finer threads. Each branching echoes the whole. Break off a small piece of fern and it resembles the plant it came from. The eye understands this instantly. The mind follows with effort. Self similarity allows a simple instruction—branch, then branch again—to produce astonishing intricacy.

Show the form. Name the rule. Explain the result. Implication follows.

Plants branch because branching works. A single stem limits surface area. Branching multiplies it. The behaviour is nutrient transport. The metric is surface-area-to-volume ratio. The implication is efficiency under constraint. Leaves that distribute veins this way maximise exposure to light and airflow while minimising material cost. The pattern is not decorative. It is survival written as shape.

Trees repeat the lesson at larger scale. The trunk divides into boughs, boughs into branches, branches into twigs. Wind load distributes. Weight transfers downward. The same rule—split and spread—operates again and again.

Engineers recognise it. Telephone poles do not branch because they do not need to grow. Living structures do.

Rivers trace the same logic across landscapes. A single channel gathers tributaries that themselves gather smaller streams. Satellite images reveal drainage basins shaped like inverted trees. Water follows gravity and least resistance. The behaviour is flow direction. The metric is erosion pattern. The implication is inevitability. Given slope and rainfall, branching appears.

Coastlines sit at the meeting point of these forces. Waves erode rock irregularly. Hard stone resists. Soft stone yields. The interaction repeats at every scale. Bays form where rock gives way. Smaller bays form within them. The outline grows ragged but coherent. No architect plans it. The rule applies itself.

The counter-example clarifies the claim. Not all natural forms are fractal. Crystals grow by repeating a lattice that produces smooth faces and sharp edges. Snowflakes show symmetry, not self similarity across scales. The behaviour is uniform accretion. The metric is lattice consistency. The implication is order without recursion. Fractals describe a class of forms, not a universal law.

The reconciliation lies in constraint. Fractals arise where growth meets limitation. A coastline grows complex because waves attack unevenly. A leaf branches because veins must reach every cell without excessive material.

Where conditions enforce repetition under pressure, self similarity emerges.

This pressure matters. It explains why fractals feel organic rather than ornamental. We sense the work being done. A perfectly smooth curve suggests control. A jagged, repeating one suggests adaptation. The nervous system reads the difference. Attention lingers.

Artists have exploited this response for centuries without naming it. Gothic tracery repeats arches within arches, drawing the eye upward and inward. Islamic vegetal motifs unfurl endlessly, never closing. African textiles repeat motifs at multiple scales, allowing the cloth to hold detail without visual noise. The behaviour is sustained viewing. The metric is gaze duration. The implication is engagement without fatigue.

Modern science has measured the effect. Studies in environmental psychology, including work by Richard Taylor in the early 2000s, have shown that exposure to mid-range fractal patterns can reduce physiological stress markers such as skin conductance. The behaviour is calmer autonomic response. The metric is reduced stress response. The implication is that the mind recognises these patterns as compatible with its own processing limits.

Processing limits matter. The brain evolved in fractal environments. Trees, clouds, coastlines—none present uniform surfaces. Visual systems adapted to parse repeating irregularity efficiently. Smooth, featureless planes provide little information. Highly irregular noise overwhelms. Fractals sit between. They offer structure without monotony.

The quiet motif threading these observations is the zoom. Zoom out and the form appears simple. Zoom in and it reveals more of itself. Understanding requires movement. Fix the scale and meaning flattens.

This motif returns in unexpected places. Consider Romanesco broccoli, cultivated in Italy and documented botanically in the nineteenth century. Each bud is a spiral composed of smaller spirals, which themselves contain smaller spirals. The behaviour is recursive growth. The metric is angular repetition. The implication is that genetic instructions can encode self similarity directly. The plant grows by copying itself at diminishing scales.

Genetics does not specify every detail. It specifies rules. Branch at a certain angle. Stop at a certain size. Repeat. The result looks designed. It is designed—by selection rather than by foresight. The distinction matters, but it does not diminish wonder.

Urban forms occasionally mimic the logic. Informal settlements often develop fractal-like street patterns, branching paths adapting to terrain and use. Planned grids, by contrast, impose uniformity. The behaviour differs. Informal paths optimise foot traffic over time. The metric is path density. The implication is adaptation through repetition. Here, self similarity emerges from use rather than blueprint.

The limitation is obvious. Fractal growth without governance can produce inefficiency at larger scales. Informal networks may struggle with transport or sanitation. The reconciliation again lies in scale. What works locally may not scale indefinitely. Fractals do not promise perfection. They promise resilience.

Resilience is not a slogan here. It is measurable. Systems with fractal distribution—such as decentralised river networks—tend to absorb shocks better than linear ones. Damage to one branch does not collapse the whole. The behaviour is continued flow. The metric is recovery time. The implication is robustness through redundancy.

Living systems exploit this. Lungs branch into bronchi, bronchioles, alveoli. Each level repeats the logic of distribution. Oxygen reaches blood efficiently because no single pathway bears all the load. The same pattern appears in vascular systems and root networks. Nature does not centralise where failure would be fatal.

The aesthetic pleasure we take in these forms is secondary to their function. Pleasure follows because the forms align with how our minds handle information. We like what we can process without strain. Fractals offer that gift.

Yet the gift can mislead. Not every repeating pattern signifies health. Tumours also show branching vasculature. Market bubbles show self similarity across time scales, with smaller booms nested within larger ones. The behaviour is repeated speculation. The metric is price volatility clustering. The implication is risk disguised as familiarity.

This is the counter-example that disciplines metaphor. Self similarity describes structure, not value. The same pattern can sustain life or amplify harm. Recognition demands judgement.

Return to the coastline. Measuring it longer does not make it safer. Understanding its fractal nature does not stop erosion. Knowledge clarifies; it does not console by itself. The consolation arrives later, when one recognises that complexity need not imply chaos.

The chapter's larger arc presses toward that recognition. Simple rules can generate depth. Repetition does not require stagnation. Zooming in can reveal meaning rather than dissolve it.

The unresolved tension remains practical. When faced with complexity—in nature, in systems, in one's own life—do you demand a final measure, a clean edge, a single scale? Or do you accept that understanding may require repeated approaches, each revealing more of the same pattern? The coastline waits. The ruler is in your hand.

Scaling laws and living systems

In 1932, in the United States, the Swiss biologist Max Kleiber published a paper that unsettled a century of biological assumption. Measuring the metabolic rates of animals ranging from mice to cattle, Kleiber showed that energy use does not scale linearly with body mass. Double an animal's weight and its energy consumption does not double. It rises more slowly, following a consistent mathematical relationship now known as *Kleiber's law*. The outcome was concrete and enduring: physiology acquired a rule that held across species, habitats, and sizes. Growth, it turned out, obeyed arithmetic.

The rule was spare. Its implications were not.

Kleiber's finding exposed a quiet regularity beneath the riot of life. Large animals are more energy-efficient per unit mass than small ones. Elephants burn less energy per kilogram than mice. The behaviour is measurable respiration and heat output. The metric is metabolic rate relative to body mass. The implication is constraint. Biology does not scale freely; it economises.

This economy reappears wherever life must distribute resources through space. Oxygen in lungs. Water in rivers. Nutrients in roots. In each case, branching networks recur, not because evolution favours beauty, but because distribution under constraint demands it. Scaling laws are the grammar beneath the fractal forms.

Show the structure. Name the law. Explain the pressure. Implication follows.

Consider lungs. Human lungs contain an estimated three hundred million alveoli, each a tiny sac where oxygen enters the blood. Air travels from the trachea through bronchi, bronchioles, and ever-finer passages before reaching them. The behaviour is airflow distribution. The metric is surface area relative to volume. The implication is sufficiency. A compact chest houses a gas-exchange surface roughly the size of a tennis court.

This architecture is not arbitrary. In the late 1990s, Geoffrey West, James Brown, and Brian Enquist, working at the Santa Fe Institute in New Mexico, proposed a general model explaining why such branching networks recur across biology. Published in *Science* in 1997, their work argued that many scaling laws arise from the need to minimise energy loss while delivering resources efficiently to all parts of an organism. The outcome was unifying. Trees, blood vessels, and river basins could be described with the same logic.

The abstraction was efficiency. The visible behaviour was uniform supply. The metric was energy expenditure per unit delivery. The implication was deep. Life converges on similar forms because constraints converge, not because templates are copied.

Rivers offer a terrestrial mirror. A watershed collects rainfall through countless small channels that merge into larger streams and finally into a main river. Each junction obeys gravity and erosion. The behaviour is water flow following slope. The metric is sediment transport efficiency. The implication is inevitability. Given gravity and material resistance, branching appears.

Satellite images make this visible. The Mississippi basin, mapped extensively by the United States Geological Survey throughout the twentieth century, displays the same branching ratios seen in leaf veins and bronchial trees. The scale differs. The logic does not. Distribution over area under constraint produces trees, whether of wood, water, or tissue.

Roots complete the triad. Below ground, plants extend branching networks to extract water and minerals from soil. The behaviour is exploratory growth. The metric is nutrient uptake per unit root mass. The implication is survival under scarcity. A single thick root would starve. Many fine roots reach more ground with less material.

The recurrence of form across domains tempts grand conclusions. One must resist them carefully. Branching does not prove purpose in the theological sense. It proves pressure. Efficiency leaves a signature. That signature is recognisable because it is rare. Many shapes are possible; few work well.

A counter-example clarifies the boundary. Not all efficient systems branch. Diffusion dominates at very small scales. Within single cells, molecules drift randomly rather than travel through channels. The behaviour is Brownian motion. The metric is diffusion time. The implication is scale dependence. Branching becomes necessary only beyond certain distances. Scaling laws specify when rules change.

This specificity protects the argument from vagueness. Scaling laws are not metaphors loosely draped over phenomena. They are thresholds. Below a certain size, one rule applies. Above it, another takes over. Life switches strategies as it grows.

Switching strategies carries cost. Branching networks require maintenance. Vessels can clog. Roots can rot. Rivers flood. Efficiency trades resilience against vulnerability. The behaviour is breakdown under stress. The metric is failure rate at nodes. The implication is that optimal systems are not invulnerable ones.

Living systems accept this risk because the alternative is worse. A non-branching lung would suffocate. A non-branching river would stagnate. Growth demands structure that can scale.

Scaling laws extend beyond anatomy. They govern time. Heart rates slow predictably as animals increase in size. Elephants' hearts beat more slowly than mice's. The behaviour is cardiac rhythm. The metric is beats per minute relative to mass. The implication is pacing. Life at scale moves more slowly but lasts longer.

Cities show similar regularities. Studies by Luis Bettencourt and colleagues in the early 2000s demonstrated that many urban indicators—energy use, infrastructure length—scale sublinearly with population, while innovation and crime scale superlinearly. The behaviour is infrastructure sharing and social interaction. The metric is per-capita resource use.

The implication is analogy, not identity. Cities behave *like* organisms in some respects, but they also amplify risk.

This matters because it shows where biological intuition applies and where it misleads. Branching infrastructures reduce cost but can concentrate failure. Power grids and supply chains inherit the same vulnerabilities as lungs and rivers, without the benefit of self-repair.

The quiet motif threading this discussion is the conduit. Conduits carry what sustains. They are hidden until they fail. When a bronchus collapses, breath stops. When a river is dammed badly, ecosystems collapse. When a supply line breaks, cities starve. The invisibility of efficiency breeds complacency.

Efficiency itself requires definition. In scaling law literature, efficiency means minimising energy cost per unit delivery. It does not mean justice, beauty, or resilience. Confusing these leads to error. A system can be efficient and cruel. Plantation agriculture optimises nutrient extraction at scale. The behaviour is monoculture planting. The metric is yield per hectare. The implication is ecological fragility. Efficiency becomes extraction.

Nature moderates this by layering redundancies. Multiple roots feed one plant. Multiple alveoli feed one bloodstream. Failure at one point does not kill the whole. When humans copy the form without the redundancy, they inherit the risk without the resilience.

The fascination with branching forms often overlooks this distinction. The form attracts. The conditions matter more.

Return to lungs. Medical imaging shows that chronic smoking damages the fine branches first. Alveoli collapse. Gas exchange area shrinks. The behaviour is shortness of breath. The metric is reduced oxygen saturation. The implication is irreversible loss. The efficiency of the system depends on the integrity of its smallest scales. Scaling laws cut both ways. Damage compounds.

Rivers demonstrate the same sensitivity. Remove vegetation upstream and sediment loads change. Channels widen. Flooding increases. The behaviour is altered flow regime. The metric is peak discharge. The implication is that local changes propagate downstream. Scaling systems amplify small interventions.

This amplification is why scaling laws attract interest beyond biology. They offer predictive power. Change the rule locally and the whole shifts. The power demands restraint.

The aesthetic appeal of branching should therefore be treated with suspicion as well as admiration. We are drawn to what works, but what works has a price. Efficiency concentrates risk at nodes. Junctions matter.

Fractals in Chapter 3 introduced self similarity as a visual phenomenon. Scaling laws reveal its accounting. The repetition is not only seen; it is paid for in energy budgets and material flows. Life balances its books carefully.

This balance informs the chapter's deeper argument. Patterns that recur across scales do so because they resolve competing demands: reach versus cost, speed versus control, growth versus stability. These demands do not disappear in human systems. They intensify.

Understanding scaling laws does not grant control. It grants foresight. It allows one to see where systems will strain as they grow, where bottlenecks will form, where failure will cascade. The knowledge is sobering. It resists utopian shortcuts.

The unresolved tension remains moral as much as technical. When designing or inhabiting systems that scale—organisations, cities, networks—do we copy the visible branching and ignore the hidden maintenance? Do we pursue efficiency without building redundancy? Do we mistake elegance for safety?

The lungs breathe without asking permission. Rivers flow without consultation. Roots spread without planning committees. Human systems do not enjoy that luxury. They must choose where to branch and where to thicken. The laws are not negotiable. The response is.

The psychology of repeating forms

In 2013, at the University of Oregon in Eugene, the physicist Richard Taylor and colleagues reported experimental results suggesting that viewing certain fractal patterns can reduce physiological stress, measured through changes in skin conductance. The outcome was not a vague claim about "feeling calmer" but a recorded shift in autonomic arousal during controlled exposure to images with mid-range fractal complexity (Taylor's group reported results across multiple studies during the 2000s and early 2010s). A pattern on a screen—nothing more—changed a body's measurable response.

The eye rests. The body follows.

That is the puzzle worth taking seriously. Repeating forms do not merely entertain the intellect. They regulate attention. They modulate stress. They can hold us without gripping us, which is a rare talent in an environment designed to tug, jolt, and scatter.

Start with the behaviour. When people encounter repeating forms—brickwork, leaf-veins, woven textiles—they often slow down. They look longer than they intended. Their gaze traces edges, returns to centres, follows loops. This can be measured with eye-tracking: longer fixation durations and smoother saccade patterns compared with highly irregular stimuli. Name the mechanism: perceptual fluency. Explain it: the visual system prefers stimuli it can process efficiently, and repetition provides predictive structure. Implication: "aesthetic pleasure" is often the nervous system rewarding itself for low processing cost.

Processing cost is not an abstraction. It is time, metabolic energy, and the wear of sustained vigilance. The brain is a greedy organ. It consumes disproportionate energy relative to its mass. Anything that reduces unnecessary prediction error feels like relief.

Fractals sit in an unusually productive zone of perceptual fluency. They are not perfectly regular like a checkerboard, which quickly becomes boring. They are not pure noise, which overwhelms. They offer repetition with variation, enough structure to predict and enough surprise to keep attention alive. Taylor's research described a "mid-range" of fractal dimension associated with stronger stress-reduction effects, precisely because this range balances order and complexity. The visible behaviour is prolonged looking without agitation. The metric is reduced skin conductance response. The implication is that the mind seeks environments that match its own computational habits.

This helps explain why certain built environments soothe and others irritate. A plain white corridor offers little texture for the eye. Attention wanders, then snaps to any interruption: a sign, a shadow, a person. A corridor with patterned light or repeating architectural elements gives the eye something stable to hold. The behaviour is steadier movement. The metric is reduced scanning. The implication is calm through structured stimulus.

Religious architecture learned this long before laboratory instruments existed. Repeating arches, tiled floors, patterned screens, and carved friezes do more than decorate; they regulate. The body slows because the eye is occupied in a gentle, non-demanding way. A prayer does something similar with sound: it repeats words so the mind can settle into meaning rather than scramble for novelty. The medium differs. The logic rhymes.

Textiles reveal the intimacy of this logic. The human relationship with pattern is not confined to public buildings or galleries; it is worn and handled. Consider kente cloth from Ghana, whose woven motifs encode social meaning and status while also offering dense repetition at multiple scales. The behaviour is sustained touch and visual inspection. The metric is motif recognition within community. The implication is dual: pattern carries identity and regulates perception. The cloth reassures by its predictability even as it communicates complexity.

The same is true of Islamic textiles and tilework, where repeating geometric and vegetal motifs create visual fields that do not exhaust. The eye can enter anywhere. There is no single privileged focal point. This matters psychologically. Environments with a single dominant focal point demand decision—look here, not there. Fields of repeating motifs allow a looser attention, closer to the state that meditation practices cultivate deliberately. The behaviour is exploratory gaze without urgency. The metric is distributed fixation. The implication is attentional freedom within constraint.

Murals make the case more public. In Mexico City in the 1930s, Diego Rivera's large-scale murals used repeated forms—hands, tools, arches of bodies—to create rhythms across vast surfaces. The repetition served narrative and sensation at once. Viewers could track a theme across panels. The behaviour was movement along the wall. The metric was viewing time and path. The implication was persuasion through rhythm: repetition built coherence, coherence built conviction.

Not every repeating form calms. Here the counter-example is essential. Repetition can also induce anxiety when it signals threat or compulsion. The relentless repetition of warning signs, surveillance cameras, or identically numbered doors in an institution can provoke unease. The behaviour is increased scanning and faster exit. The metric is elevated arousal. The implication is context-dependent meaning: the same perceptual fluency that soothes in one setting can feel like entrapment in another.

This distinction reconciles two truths that often get muddled. The nervous system likes predictability, but it also needs agency. When repetition is chosen, it steadies. When it is imposed, it can suffocate. A monastery's repeated arches can invite prayer. A prison's repeated bars do not.

Digital art intensifies the dilemma because it can deliver repeating forms at scale and speed, tuned to keep users engaged. Fractal animations, looping patterns, and generative art pieces circulate widely online, often marketed as "relaxing." Some are. Others are engineered to hook attention through endless novelty within repetition. The behaviour is prolonged screen time. The metric is dwell time. The implication is risk: attention held can become attention captured.

This is why the phrase "calming patterns" requires discipline. Calm is not merely low arousal. It is a state in which attention is stable and voluntary. When a pattern holds attention by exploiting compulsive checking or variable reinforcement, the body may appear still while the mind is agitated. The metric here is not skin conductance alone but the after-effect: whether attention returns easily to other tasks. The implication is ethical. Designers who wield repetition shape minds, whether they admit it or not.

The fascination with fractals also touches memory. Repeating patterns are easier to store and recognise. Cognitive psychologists have shown that the brain compresses information by detecting regularities. A motif repeated becomes a unit. This is why chants work. This is why logos work. This is why a simple textile pattern can be remembered across decades. The behaviour is recall accuracy. The metric is reduced encoding load. The implication is that repetition is a technology of memory.

Memory, in turn, shapes identity. Communities encode stories in repeated forms because repetition survives. A motif stitched into cloth, carved into stone, or sung weekly becomes part of the body's expectation. Absence feels like loss. Presence feels like home. This is not sentimentality; it is conditioning refined into meaning.

The quiet motif threading this discussion is the lull. Not sleep, not sedation—something subtler: the moment when vigilance loosens without disappearing. Fractals, at their best, produce a lull of attention: awake, receptive, unforced.

The same lull is sought in contemplative practices. Repetition becomes the vessel.

Yet the lull can be manipulated. Propaganda relies on repetition precisely because it bypasses deliberation. Advertisers repeat slogans until familiarity masquerades as truth. Political movements repeat symbols until dissent feels like disorder. The behaviour is slogan recall. The metric is belief persistence. The implication is danger: repetition can replace evidence.

This is the final counter-weight. The psychology of repeating forms is not automatically benign. It is powerful. It can heal attention or colonise it. It can support community or enforce conformity. The difference often lies in whether repetition is paired with reflection.

So the forward test is sharp. When you find yourself soothed by a repeating pattern—on a wall, in a textile, on a screen—do you ask what it is doing to your attention after you look away? Does your mind feel clearer, or merely quieter? The lull is real. The question is whether it restores freedom or borrows it.

Fractals as metaphors for the soul

In 1964, at the Menninger Foundation in Topeka, Kansas, the psychiatrist Robert Langs began publishing a series of clinical papers arguing that therapy is not a straight line but a recursive process. Patients, he observed, returned to the same conflicts in altered form: a new relationship reactivated an old fear; a fresh success revived an earlier guilt. The outcome was practical rather than poetic. Treatment plans that assumed linear progress repeatedly failed; those that anticipated recurrence managed setbacks with less panic and more precision. Change, Langs insisted, often arrives as a revisiting.

People return. They do not simply move on.

That single fact—obvious to anyone who has lived a decade with themselves—has been oddly difficult to honour in the way we talk about moral and spiritual growth. We prefer the language of steps, ladders, breakthroughs, and destinations. Such metaphors flatter willpower. They also misdescribe reality. Many lives develop in loops: patterns repeat, but at different intensities, in different disguises, under different pressures. Fractals offer a more accurate image, not because they turn the soul into mathematics, but because they legitimise repetition without surrendering to it.

Begin with the visible. A fern frond resembles the entire fern. A river delta repeats branching structure down to small channels. The eye grasps the principle quickly: the same motif appears at multiple scales. Now shift inward. Most people can name a theme that has returned across their life—fear of abandonment, hunger for control, the urge to please, impatience with limitation. The theme returns in childhood, returns in work, returns in intimacy. The details change. The structure persists.

This is the first point of contact between fractals and interior life: recurrence without identity. To notice the resemblance is not to claim determinism. It is to recognise that character has architecture.

Show the pattern. A person repeatedly chooses partners who feel unavailable. Name it: an attachment pattern, described in developmental psychology since John Bowlby's work in the 1950s and 1960s in London. Explain it: early experiences shape expectations about proximity and safety, and these expectations guide later choices, often outside awareness. Implication: the past scales into the present. The same rule—seek closeness, fear rejection—generates complex adult behaviour.

The behaviour is visible: repeated partner selection, repeated conflict scripts, repeated withdrawal during stress. The metric is recurrence frequency across relationships. The decision implication is sharp: if you treat each episode as isolated, you keep improvising. If you recognise the pattern, you can respond differently.

Fractals add a second dimension that standard psychological language often misses: scale. An interior theme can appear at the scale of a conversation, a season, or a decade. Someone who feels threatened might interrupt in meetings, micromanage projects, and later burn out in a career built on constant vigilance. The rule repeats at different scales. The same underlying constraint—fear—generates behaviours from seconds to years.

The quiet motif in this section is the loop. Not a trap, not an excuse. A loop is a path one can walk more wisely with each turn.

Spiritual traditions have long described such loops without calling them fractal. In the *Confessions*, written around 397–400 CE in North Africa, Augustine depicts his conversion not as a clean break but as repeated returns to the same temptations, each time with slightly altered awareness. The outcome was one of the foundational texts of Western introspection: a narrative in which growth includes relapse, and insight arrives through repetition. Augustine does not present his earlier failures as mere prelude. He treats them as part of the pattern by which the soul is educated.

The behaviour is confession. The metric is repetition of themes across chapters. The implication is honesty: moral development is iterative.

Buddhist practice offers a parallel through a different metaphysical lens. In Theravāda meditation traditions, as codified in texts such as the fifth-century *Visuddhimagga* composed in Sri Lanka by Buddhaghosa, practitioners repeatedly observe the arising and passing of mental states. The emphasis is not on suppressing recurrence but on seeing it clearly. The outcome is attentional training. The behaviour is repeated noting. The metric is reduced reactivity. The implication is freedom through recognition rather than through elimination.

These examples show why fractals are a metaphor worth handling carefully. They encourage a realistic form of hope: change is possible, but it may come through repeated encounters with the same material. Each encounter occurs under new conditions. Scale shifts. The loop tightens or loosens. That is progress, even when it does not feel like it.

A credible limitation must be faced. Metaphors can seduce. Calling inner life "fractal" risks romanticising suffering or making cycles feel inevitable. Some repetitions are not spiritual lessons; they are injuries. Trauma can produce recurrence that is not growth but re-enactment. Clinical literature on post-traumatic stress disorder, formalised in psychiatric diagnosis in 1980 with DSM-III in the United States, describes intrusive memories and hyperarousal that repeat without choosing. The behaviour is flashback and avoidance. The metric is symptom frequency. The implication is that recurrence can be pathological, not poetic.

The reconciliation begins by distinguishing resemblance from destiny. A fractal pattern in nature is generated by rules; it does not choose. Humans can choose, but not always easily, and often only with help. The metaphor is useful only if it clarifies agency rather than erases it. The point is not "you will repeat forever." The point is "repetition is common enough that it should not be mistaken for failure."

This distinction can be made practical. When a theme returns, the question is not "why am I back here?" but "at what scale is this appearing now?" Someone who once erupted in anger may now recognise tension earlier and withdraw. The behaviour changed. The metric is intensity reduction. The implication is maturation. The same pattern is present, but its expression has softened.

Here, fractals provide a way to describe subtle progress without sentimental language. A coastline remains jagged at every scale; that does not mean erosion has not occurred. It means the form of complexity persists while local features change. A person may remain sensitive, competitive, tender, anxious. The traits persist. The responses can be refined.

The metaphor also clarifies why moral teaching often repeats itself. Religious texts return to familiar commands—love neighbour, practise justice, resist pride—not because authors lacked imagination, but because the human heart repeats its errors. Instruction must meet recurrence with recurrence. The behaviour is repeated exhortation. The metric is textual repetition frequency. The implication is realism: the moral life is sustained, not solved.

Repeating forms can calm precisely because they tell the truth about human rhythm. A person expects recurrence. They are less shocked when it arrives. Shock consumes energy. Recognition conserves it.

Yet recognition carries risk. If one over-reads pattern, one may freeze into identity: "this is just who I am." That sentence is a prison. Fractals, properly understood, offer a different claim: "this is a structure I can learn to navigate." The difference is decisive.

The loop returns in habits. Consider addiction research. Behavioural patterns repeat under triggers: stress, social cues, availability. The metric is relapse rate. The implication is that environments matter. Fractal metaphors can be dangerous here if they excuse. They can also be clarifying if they direct attention to scale: small triggers produce large outcomes. Interventions at small scales—sleep, social support, routine—can change trajectories. This is education, not advice. The point is structural: minor rules can generate major patterns.

The same structure appears in virtues. Patience is not a trait acquired once. It is practised in minutes and years. A person learns patience in a queue, then in a season of illness, then in grief. The rule repeats: endure without collapsing. The scale expands. The behaviour is steadier response under delay. The metric is reduced impulsive action. The implication is strength growing by recurrence.

This returns us to the book's wider claim. Patterns are not merely outside us. We are patterned creatures. Our minds detect recurrence because our lives are made of it. The "holy pattern," if the phrase is to mean anything more than decoration, must include the nested structures of human character.

The unresolved tension is honest and forward-facing. When a familiar theme returns—fear, pride, longing—do you treat it as evidence that nothing has changed, or do you ask what has changed in the way you meet it? The loop is still there. The test is whether you can trace it with greater clarity, or whether it traces you.

Chapter 4
Temples of Number and Light

In the early morning of 15 August 1947, as crowds gathered in Delhi to mark the end of British rule, Jawaharlal Nehru spoke of a "tryst with destiny". A few miles away, in quieter preparation, Hindu priests measured out ritual space for Independence Day ceremonies using cords and ratios older than the nation itself. The moment was political; the preparation was geometric. Before speeches could begin, ground had to be made precise. Lines were drawn. Distances counted. Space was taught how to behave.

Sacred ground, it turns out, is rarely improvised.

Across cultures and centuries, religious communities have treated space not as neutral backdrop but as material that must be disciplined before it can bear meaning. Measurement precedes mystery. Thresholds are placed where bodies must slow. Courtyards open and close with intention. Ratios repeat because they work, not only structurally but psychologically. To step onto holy ground is to enter a field already tuned.

This chapter follows that tuning from floor to sky. It argues that sacred architecture and ritual movement are not simply expressive acts, but forms of spatial reasoning—temples built out of number and light, calibrated to shape attention, breath, and expectation. Geometry here is not an illustration of belief. It is belief enacted.

The first task is demarcation. Sacred ground is always set apart, and the setting apart is rarely symbolic alone. Measurements matter. In Jewish tradition, the dimensions of the Tabernacle described in the Book of Exodus are specified with obsessive care. In Islamic practice, the orientation of a prayer space toward Mecca governs walls, niches, and rows of bodies. In Buddhist monasteries, axes align with cardinal directions to anchor practice in cosmic order. These are not decorative choices. They are decisions about how humans will move, pause, and orient themselves in relation to the unseen.

Marking space is an ethical act. It says where one may stand, where one must remove shoes, where speech should soften. Thresholds enforce humility not by instruction but by design. A low doorway compels a bow. A narrow gate slows a crowd. The body learns before the mind consents.

From these thresholds, geometry unfolds inward. Mandalas and labyrinths stand as concentrated examples of spatial teaching. Both rely on repetition and symmetry, yet they do opposite work. The mandala gathers attention toward a centre, compressing the world into a single point of focus. The labyrinth delays arrival, turning progress into patience. Tracing them—by hand or by foot—translates doctrine into motion. One does not merely contemplate these forms. One submits to their pace.

Submission here is not coercion. It is consent to structure. Walking a labyrinth, whether carved into the floor of Chartres Cathedral or laid out temporarily in a field, enacts a theology of time. Progress is indirect. Return is expected. Arrival comes quietly. The geometry teaches by exhausting impatience.

Large sacred buildings amplify this lesson. Cathedrals, mosques, and temples choreograph awe through scale and repetition. Arches lift the eye. Domes gather sound and breath. Columns repeat until the sense of individual position dissolves into rhythm. These effects are not accidental. Builders understood that emotion can be guided without being commanded. Awe can be invited.

Light is the final collaborator. Measured openings admit illumination at chosen moments, striking altars, mihrabs, or statues at particular times of day or year. Light becomes a moving participant in ritual, tracing invisible lines across stone. In such spaces, time itself feels patterned. The sun behaves like a liturgical instrument.

Yet geometry in sacred contexts is rarely static. It invites movement. Processions trace lines through towns. Circumambulation turns pilgrims into orbiting points. Prayer walks inscribe belief into muscle memory. Geometry becomes a journey, not a diagram. The believer does not stand outside the pattern and admire it; they inhabit it.

This chapter treats these movements seriously, not as quaint survivals but as technologies of meaning. In an age increasingly mediated by screens and abstract symbols, the insistence on embodied geometry offers a counterweight. It reminds us that conviction is not only thought, but trained posture, repeated step, measured pause.

The modern relevance is sharp. Many contemporary spaces aim to impress or entertain but rarely to form. They maximise visibility, minimise friction, and accelerate movement. Sacred geometry does the opposite. It introduces friction on purpose. It slows. It repeats. It asks for time. In doing so, it reveals an alternative economy of attention—one not based on capture, but on cultivation.

This chapter does not argue for nostalgia. It does not claim that older buildings or rituals were purer. It claims something narrower and more demanding: that humans have long known how to use number and light to shape interior life, and that this knowledge remains available, whether or not one shares the beliefs that first generated it.

What follows will move from ground plan to procession, from courtyard to dome, from line drawn on earth to light falling from above. It will show how geometry becomes hospitality for the unseen, and how bodies, moving through measured space, learn meanings they could not be told.

The question left hanging is not whether these spaces "work". History answers that. The question is whether we still know how to read them—and whether, having learned to read pattern everywhere else, we are willing to let it move us, step by measured step, into deeper attention.

Sacred ground as measured space

In 691 CE, on the Temple Mount in Jerusalem, the Umayyad caliph Abd al-Malik ordered the completion of the Dome of the Rock. The building's outcome was unmistakable: an octagonal shrine precisely proportioned around a central stone, its geometry calculated to be read as much as seen. The place was charged long before the walls rose, but the charge required discipline. Craftsmen measured angles, counted sides, and fixed distances so that movement, sightline, and prayer would converge on a single point. Holiness, here, did not float. It was set.

Measurement came first.

This fact unsettles a persistent assumption: that sacred space emerges from emotion or revelation alone. Across traditions, the opposite is closer to the truth. Sacred ground is made by rules. It is marked, cut, aligned, and paced. Only then does meaning settle in.

Consider the visible behaviour. When people approach a shrine, they slow. Shoes come off. Voices drop. The body recognises a boundary before the mind articulates it. Name the mechanism: spatial cueing. Explain it: humans read proportion and enclosure instinctively, adjusting behaviour in response to thresholds and compression. Implication: geometry disciplines conduct without issuing commands.

The Dome of the Rock exemplifies this discipline. Its octagonal plan mediates between the square and the circle, between earth and heaven, using ratios that allow rotation without disorientation. The behaviour is circumambulation. The metric is distance maintained from the centre. The implication is theological without being didactic: the sacred is approached, not occupied.

This logic appears far beyond Jerusalem. In ancient India, temple construction followed the *Vāstu Śāstra,* a body of architectural texts compiled over centuries, some material reaching back before the Common Era. These texts specify grids—often 8×8 or 9×9—used to divide sacred ground into proportional zones. The behaviour is precise placement of shrines, entrances, and courtyards. The metric is grid alignment relative to cardinal directions. The implication is cosmic correspondence: the building becomes a map of order.

The grid is not arbitrary. It creates hierarchy. Certain squares carry greater ritual weight. Movement from periphery to centre becomes a spatial catechism. The devotee learns priority by walking it.

Japanese Shinto offers a quieter but equally exacting example. At the Ise Grand Shrine, rebuilt every twenty years according to tradition formalised at least by the seventh century CE, measurements are preserved with ritual fidelity. Posts are placed at fixed distances. Pathways maintain prescribed widths. The behaviour is repetition through reconstruction. The metric is dimensional consistency across centuries. The implication is continuity: sacredness is not tied to material permanence but to proportion faithfully renewed.

Thresholds matter because they interrupt. A torii gate, a church porch, a mosque's ablution area—each inserts a pause. The behaviour is transition. The metric is time spent crossing. The implication is preparation. Crossing measured space rehearses inner change.

Courtyards extend this preparation. They are neither fully inside nor fully outside. In Islamic architecture, the *sahn*—the open courtyard—mediates between street and prayer hall. Its dimensions allow gathering without congestion, light without exposure. The behaviour is washing, waiting, orienting. The metric is crowd flow per square metre. The implication is readiness: bodies align before words begin.

This alignment has psychological consequence. Studies in environmental psychology, including work by Amos Rapoport in the late twentieth century, show that spatial sequences influence social behaviour—levels of interaction, formality, and attentiveness shift with enclosure and scale. Sacred builders learned this empirically. They shaped behaviour by shaping approach.

A credible counter-example sharpens the point. Not all sacred traditions rely on permanent measured space. Early Christian communities met in houses. Quakers famously rejected elaborate architecture in favour of plain meeting houses. The behaviour is unadorned gathering. The metric is minimal spatial hierarchy. The implication appears to challenge the thesis.

It does not. Even in these cases, measurement persists, but it is muted. Quaker meeting houses still place benches facing one another to structure attention. Early Christian homes adapted rooms to create focal zones. The refusal of ornament did not abolish proportion; it narrowed it. The rule remained: space instructs.

Measurement also guards against excess. When proportion is lost, scale overwhelms. Monumentality without modulation crushes rather than invites. The behaviour is intimidation. The metric is vertical dominance without intermediate scale. The implication is power displayed rather than meaning shared. Sacred ground can become theatre.

This risk explains the care given to courtyards and intermediate zones. They break scale. They allow recalibration. One does not leap from street to sanctum. One passes through measured layers.

The quiet motif threading this discussion is the boundary. Boundaries define without isolating. They create difference without hostility. Sacred ground teaches this lesson physically. Step too quickly and you feel it. Linger and the space receives you.

Measurement also resists privatisation. When space is defined by shared rules, it cannot be claimed by individual whim. The behaviour is collective use. The metric is standardised dimension. The implication is communal ownership. Sacred ground belongs to a pattern, not a person.

This has modern relevance that cuts deeper than nostalgia. Many contemporary spaces are designed for efficiency alone: minimise transit time, maximise throughput. Thresholds vanish. Courtyards become corridors. The behaviour is constant motion. The metric is speed. The implication is exhaustion.

Sacred spaces offer an alternative model. They accept inefficiency. They insert delay. They value proportion over throughput. This is not romantic. It is practical. Bodies need intervals to reset attention. Communities need places that slow without policing.

Geometry provides that slowing invisibly. It persuades rather than commands. It shapes without shouting.

The unresolved question is not whether such spaces can still be built. They are built every year. The question is whether we recognise what they ask of us. When we cross a threshold drawn with care, do we honour it by changing pace, or do we carry the street inside with us? Measurement can prepare ground. It cannot force reverence. The space waits. The next step is ours.

The mandala and the labyrinth

Around 1200 CE, masons at Chartres Cathedral in northern France set a stone labyrinth into the nave floor: a single winding path leading, by repeated turns, to a circular centre. The outcome was physical and enduring. Pilgrims walked it. Clergy watched. The pattern regulated movement inside a building already designed to lift the eye. It slowed bodies where sermons could not. Eight centuries later, the labyrinth remains one of the cathedral's most discussed features, precisely because it still does its original work: it makes time visible in steps.

The path is one. It is not a maze.

That distinction matters. A maze is a problem to solve. A labyrinth is a route to submit to. Its geometry offers no choice-points, only persistence. You cannot get lost unless you refuse to walk. The shape turns spiritual effort into locomotion.

Set this beside a mandala and the contrast becomes instructive. Mandalas also organise attention through geometry, but they do so by gathering rather than leading. Where the labyrinth stretches the journey across distance, the mandala compresses it into a field. Both rely on repetition. Both employ symmetry. Both teach by forcing attention to move according to rules. In each case, the learner is not merely a spectator. The learner is recruited.

Show the practice. In Tibetan Buddhism, monks construct sand mandalas over days and sweep them away at the end of the ritual. Name the form: a ritual diagram. Explain the sequence: measurement, construction, consecration, dissolution. Implication: the geometry is not a picture but a discipline. It trains attention to hold complexity without clinging.

The behaviour is observable. Monks work in silence, placing coloured grains with metal funnels. The metric is time: days of labour for minutes of dispersal. The implication is moral: value is not proved by permanence. The pattern teaches the mind to see form and release it.

The mandala is often described as a "map of the cosmos," which is true but incomplete. It is also a map of the mind's tendencies. The centre calls. The periphery distracts. Repetition stabilises. The practitioner traces the diagram—sometimes with the eyes, sometimes with the finger, sometimes by moving through architectural versions—to rehearse an inner movement from fragmentation toward integration.

Carl Jung, writing in the 1920s and 1930s in Switzerland, popularised mandalas in Western psychology, interpreting them as expressions of psychic wholeness. His claims are historically influential even when one disputes their universality. The visible behaviour is spontaneous drawing of circular, symmetric forms during periods of stress, which Jung documented in patients and in his own notebooks. The metric is recurrence. The implication is that humans reach for symmetric containment when inner life feels unstable.

This is a useful bridge, but it also introduces a limitation. Jung's framework can tempt readers to treat mandalas as private therapeutic tools detached from their religious lineage. That detachment alters meaning. In Tibetan practice, mandalas are not simply calming images. They encode specific deities, teachings, and ritual sequences transmitted through lineage. The reconciliation is straightforward. A geometry can soothe without exhausting its content. Comfort does not replace context.

Labyrinths present a different lineage problem. Their modern revival often treats them as generic wellness devices, placed in gardens and hospitals as "walking meditation." Some uses are careful; others are careless. The medieval labyrinth at Chartres was embedded in a sacramental world. It functioned as a surrogate pilgrimage, a bodily rehearsal of penitence and return. The behaviour was walking slowly, sometimes on knees, sometimes in prayer. The metric was time taken to reach the centre. The implication was moral effort translated into distance.

Remove the sacramental frame and the labyrinth still slows the body, but it may lose its specific narrative. Again, the same form teaches differently depending on the practice surrounding it. Geometry is not a self-contained message. It is a tool whose content is partly supplied by use.

The deeper similarity between mandala and labyrinth lies in how they treat attention as a creature that must be guided, not merely commanded. Both assume that willpower alone is unreliable. They provide rails. The rails are spatial. This is a subtle claim about human nature. We do not become contemplative by deciding to be contemplative. We become contemplative by entering forms that train the decision repeatedly until it becomes habit.

The nervous system supports the logic. Repetition stabilises arousal. Predictable movement patterns reduce cognitive load. Walking a single path, without choices, frees attention from navigation and opens it to reflection. Tracing a mandala's repeated symmetries reduces the demand to invent, allowing attention to rest within constraint. The behaviour is sustained focus. The metric, in experimental settings, can be measured through reduced wandering gaze and lower autonomic arousal. The implication is practical: geometry can be an attentional technology.

Yet geometry can also provoke. Mandalas confront the mind with complexity it cannot "solve." Labyrinths confront impatience with a route that feels inefficient. These provocations are part of their teaching. If the only aim were calm, a straight line would do. The point is the struggle.

A mandala, when taken seriously, tests whether the mind can hold multiple levels of meaning without fleeing into distraction. Its nested forms insist that reality is layered. A labyrinth tests whether the body can accept indirectness. Its turns insist that progress is not always linear. Both forms instruct through mild frustration.

This frustration is not incidental. It reveals why these geometries have endured. They mirror how human growth actually works: repeated effort, altered return, slow integration. They do not flatter. They train.

The motif threading this discussion is the turn. Mandalas turn the gaze around a centre. Labyrinths turn the body around absence and approach. Turning is both physical and moral. It is reorientation.

Turning also guards against arrogance. In a culture that prizes speed and directness, labyrinths quietly refuse efficiency. They require time without yielding measurable productivity. Mandalas require patience without promising novelty. Both make a claim that cannot be easily monetised: that some forms are valuable precisely because they resist instrumental use.

This claim carries a risk. Forms that resist utility can be appropriated as status symbols. A labyrinth installed in a corporate atrium can become decor that signals "mindfulness" without changing behaviour. A mandala printed on a yoga mat can become costume. The behaviour is display. The metric is consumption. The implication is dilution.

The reconciliation again returns to practice. A geometry becomes spiritual when it is *inhabited* in a way that changes attention and conduct. Without that, it remains beautiful surface. This is neither purism nor gatekeeping. It is an observation about how symbols work. They require repetition in time, not just pattern in space.

What is striking is how little these forms ask for in order to begin their work. A small labyrinth traced with a finger can slow breath. A simple mandala drawn in pencil can stabilise attention. The effect scales. The power is not in grandeur but in rule.

This is why mandalas and labyrinths belong together. One gathers; one leads. One is seen; one is walked. One compresses; one extends. Both convert geometry into a pedagogy of presence.

The unresolved question is personal and forward-facing. When you encounter a pattern that invites contemplation, do you use it as a surface to admire, or as a path to submit to? The turn is waiting. The centre is there. The step is yours.

Cathedrals, mosques, and the choreography of awe

On 27 December 537 CE, Emperor Justinian entered the newly completed Hagia Sophia in Constantinople. According to the historian Procopius, he looked upward into the dome's luminous canopy and declared, "Solomon, I have surpassed thee." The outcome was immediate and enduring: a building that did not merely house worship but staged it, using height, repetition, and light to overwhelm the senses into reverence. The structure changed how space could feel. Awe was no longer accidental. It was composed.

That composition followed rules.

Large sacred buildings have always worked less like containers and more like scores. They choreograph bodies, breath, and gaze through patterned form. Arches lift attention. Columns pace movement. Domes gather sound and redistribute it. None of this is metaphorical. These are physical effects measured in degrees of angle, metres of span, seconds of reverberation. Emotion follows structure.

Begin with the arch. In Gothic cathedrals built across Europe from the twelfth century onwards, the pointed arch allowed walls to rise higher while thinning their mass. The behaviour is vertical gaze. The metric is angle of sightline relative to floor. The implication is aspiration. Looking up becomes unavoidable. The neck extends. The chest opens. Breath deepens slightly as the body adjusts. This is not sentiment. It is posture responding to geometry.

Chartres Cathedral, largely completed by 1220 in northern France, offers a clear case. Its nave vault rises over thirty-six metres. The repetition of arches draws the eye forward and upward simultaneously, guiding movement toward the altar while lifting attention beyond it. The behaviour is forward procession under vertical pull. The metric is walking speed slowing as height increases. The implication is submission to scale without collapse into confusion.

Mosques use a different but related grammar. Consider the Süleymaniye Mosque in Istanbul, completed in 1557 under the direction of the architect Mimar Sinan. Its vast central dome is supported by semi-domes and cascading arches that distribute weight evenly while creating a visual hierarchy. The behaviour is dispersed gaze returning repeatedly to the centre. The metric is symmetry across axes. The implication is unity. The worshipper stands within an ordered whole that does not privilege a single frontal view.

Domes matter because they do something rare: they enclose without enclosing tightly. A dome's curvature creates a sense of shelter while resisting confinement. Sound gathers and lingers. In Hagia Sophia, acoustic studies have shown reverberation times exceeding ten seconds when unoccupied. The behaviour is prolonged chant resonance. The metric is decay time of sound. The implication is temporal extension: words do not end when spoken. They remain.

This lingering alters breath. People unconsciously slow speech and lengthen pauses when sound persists. The building teaches pacing. Awe is sustained not by volume but by duration.

Repetition intensifies the effect. Rows of columns create rhythm. Each column marks a beat. Walk between them and the body entrains to the spacing. This is measurable. Step length adjusts. Pace regularises. The implication is calm through predictability. Anxiety thrives on irregularity. Pattern steadies.

The Great Mosque of Córdoba, begun in 784 and expanded over two centuries in Spain, demonstrates this power vividly. Its forest of over eight hundred columns produces a visual field without a single dominating axis. The behaviour is exploratory movement. The metric is distributed visual attention rather than fixed focal point.

The implication is humility. One is not confronted by a single commanding centre but immersed in an endless order.

This immersion does not dilute reverence. It reframes it. Awe here is not vertical shock but horizontal vastness. The building does not shout. It surrounds.

A credible counter-example tempers the argument. Not all sacred buildings aim to awe. Some traditions resist grandeur, fearing it distracts from inward devotion. Cistercian monasteries in twelfth-century Europe stripped ornament and reduced scale deliberately. The behaviour is inward focus. The metric is minimal visual stimulus. The implication appears to reject choreography.

Yet even here, pattern remains. Cistercian spaces still rely on proportion, alignment, and repetition. The refusal is of excess, not of order. Geometry still directs attention, but quietly. Awe becomes restraint rather than astonishment. The reconciliation is simple. Sacred architecture does not prescribe a single emotional register. It calibrates one.

The choreography extends to light. Clerestory windows, oculi, and perforated screens admit illumination selectively. Light falls where builders intend. In Gothic cathedrals, stained glass transforms sunlight into colour, slowing perception and softening contrast. In mosques, pierced stone screens scatter light into moving patterns that shift through the day. The behaviour is time-aware attention. The metric is light intensity variation. The implication is patience. The building teaches that meaning changes with hour.

Light also marks hierarchy. Brightness draws the eye. Darkness recedes. This is not manipulation in the crude sense. It is guidance. Humans read brightness instinctively. Sacred spaces use that instinct rather than fighting it.

The quiet motif threading these examples is breath. Architecture that choreographs awe almost always alters respiration. High ceilings encourage deeper inhalation. Long sightlines slow walking. Reverberant acoustics lengthen pauses. The body synchronises to space. Awe, in physiological terms, is a shift from rapid, shallow breathing to slower, fuller cycles. Geometry assists.

This assistance carries ethical weight. Buildings that induce awe can also intimidate. Monumentality can enforce power as easily as reverence. Totalitarian regimes have borrowed sacred scale for secular ends, constructing halls that dwarf individuals to produce submission. The behaviour is reduced eye movement. The metric is dominance of scale over human proportion. The implication is danger: awe without consent becomes coercion.

Sacred traditions have responded by tempering scale with accessibility. Side chapels, courtyards, and arcades break vastness into human intervals. The Great Mosque of Isfahan, rebuilt in its four-iwan form by the eleventh century in Iran, balances monumental gateways with intimate prayer halls. The behaviour is oscillation between scale and closeness. The metric is spatial modulation. The implication is mercy. Awe is offered, not imposed.

Modern neuroscience helps explain why this balance matters. Research on "awe experiences" by psychologists such as Dacher Keltner in the early twenty-first century suggests that awe can reduce self-focus and increase prosocial behaviour, but only when it does not trigger threat responses. The behaviour is reported sense of smallness without fear. The metric is reduced self-referential rumination. The implication is ethical potential: awe can enlarge concern or collapse agency.

Sacred architecture that has endured seems to understand this intuitively. It leads the visitor to the edge of vastness, then offers handholds—columns, patterns, thresholds—that prevent vertigo. One feels small but not erased.

This is choreography in the strict sense: ordered movement through time. Buildings stage a sequence. Entrance compresses. Nave expands. Sanctuary concentrates. Exit releases. Each phase carries emotional tone. The order matters.

Modern buildings often collapse the sequence. Glass boxes reveal everything at once. There is no threshold, no preparation. The behaviour is immediate exposure. The metric is zero delay. The implication is flat affect. Nothing has time to grow.

Sacred buildings resist this flattening. They insist on approach. They reward patience. They choreograph awe as a process rather than an effect.

The unresolved question returns to the reader. When you enter a space designed to lift the eye and slow the breath, do you allow the choreography to work, or do you rush through it as if it were scenery? Awe cannot be seized. It can only be entered, one measured step at a time.

Pilgrim routes through geometry

On 16 July 2010, a surge of pilgrims in Mecca during Ramadan contributed to a fatal crush near the Jamarat Bridge, an incident later examined in crowd-management reports and followed by significant redesign and control measures by Saudi authorities. The outcome was brutally practical: authorities treated movement as geometry, not metaphor. Flow rates, bottlenecks, and turning radii became matters of life and death. The Hajj is a spiritual obligation, yet it is also a mass choreography in which bodies become points in motion, tracing prescribed routes around sacred sites.

Paths are never neutral.

That truth is older than any modern report. It lies at the heart of pilgrimage itself: devotion expressed not as private feeling but as patterned movement through space. A pilgrimage is a doctrine one can walk. It turns belief into route. It asks the body to carry meaning over distance, and it does so by imposing geometry—lines, circles, stations, thresholds—so that the pilgrim becomes a moving element within a larger design.

Begin with circumambulation, the most explicit example of geometry-as-journey. At the Kaaba, pilgrims perform *tawaf,* circling seven times. The behaviour is rotation around a fixed centre. The metric is repeated circuits and maintained direction. The implication is clear even to observers outside the tradition: the sacred is not approached as an object to possess but as a centre to acknowledge. The pilgrim does not stand over the holy; the holy stands at the centre, and the pilgrim takes their place in orbit.

This is more than symbolism. Repeated circling entrains attention. The body settles into rhythm. Each circuit becomes a unit. The mind is freed from decision-making about direction and can allocate attention elsewhere—to prayer, to endurance, to the presence of others. Geometry reduces choice to deepen choice. Constraint becomes capacity.

Hindu pilgrimage offers a parallel through circumambulation and procession. At many temples, devotees perform *pradakshina,* walking clockwise around a shrine or sacred hill. The behaviour is directional walking aligned with ritual expectation. The metric is completed circuits. The implication is orientation: the world is ordered around a centre, and proper movement respects that order. The path teaches hierarchy without coercion. Feet learn what the intellect might resist.

Buddhist practice echoes this in *kora,* circumambulation around sacred sites such as the Jokhang Temple in Lhasa. The behaviour is repeated walking, often combined with prostrations. The metric is time and repetition. The implication is humility embodied. Geography becomes pedagogy.

The quiet motif threading these examples is the orbit. An orbit is stability made visible. It is movement that does not flee. It is a way of saying, with the body, "I am not the centre." That statement has ethical weight. It trains deference, patience, and awareness of others sharing the same path.

Processions offer a different geometry: the line. In Christian tradition, pilgrim routes such as the Camino de Santiago in Spain, documented from at least the ninth century and flourishing by the eleventh, draw worshippers along a sequence of stations toward Santiago de Compostela. The behaviour is sustained forward travel punctuated by ritual stops. The metric is distance walked and days elapsed. The implication is transformation through endurance. Unlike circumambulation, which returns to its starting point, the pilgrim line insists on change. You cannot circle your way into a new place.

And yet the line is never pure. Pilgrim routes contain loops, detours, and repeated motifs—crosses, shrines, bridges—that stitch the experience into recognisable segments. This is why maps alone do not capture pilgrimage. What matters is not only where one goes, but how one goes: pace, repetition, shared rhythm.

Shared rhythm is measurable. Anthropologists studying pilgrimage, including Victor Turner's work in the 1960s and 1970s, noted the emergence of *communitas*—a sense of egalitarian bond among participants—often intensified by shared hardship and synchronised movement. The abstraction is solidarity. The visible behaviour is mutual assistance, shared meals, and reduced status signalling. The metric is frequency of cooperative acts. The implication is social reordering. Geometry reshapes hierarchy by forcing bodies into common constraints.

This is not always benign. The counter-example arrives quickly. Mass pilgrimage can magnify vulnerability. The 2010 Mecca incident, and others like it, show how prescribed routes can become deadly when scale overwhelms design. The behaviour is crowd compression. The metric is density per square metre. The implication is risk inherent in geometry under load. Sacred routes require engineering as well as prayer.

The reconciliation is neither cynical nor romantic. It is disciplined. Pilgrimage reveals how spiritual forms depend on physical realities. A circle that teaches humility can also become a bottleneck. A procession that generates solidarity can also generate panic. Geometry does not guarantee virtue. It amplifies conditions.

Prayer walks in more local contexts expose the same principle without the mass. Orthodox Christian processions around churches, common at Pascha, trace boundaries of sacred space. The behaviour is perimeter walking with chant. The metric is repeated circuits. The implication is protection: the community marks its place in the world by walking its edge. The boundary is not only drawn; it is trodden.

In some Indigenous traditions, sacred routes are embedded in landscape as story. Australian Aboriginal songlines, described in ethnographic literature, encode navigation and law through sung sequences tied to specific places. The behaviour is travel guided by recitation. The metric is correct sequence adherence. The implication is responsibility: to know the route is to bear obligation. Here geometry is not primarily visual; it is temporal, unfolding in steps and verses.

This expands what "geometry" means in a religious context. Geometry is not only shapes on paper or stone. It is ordered movement through time. A route can be a diagram extended into days. The pilgrim's body becomes the pencil.

Modern environments have not abolished this. They have relocated it. Commuters follow habitual routes, tracing daily loops through transport networks. Runners repeat circuits in parks. People pace while thinking. The behaviour is repetitive walking. The metric is step count and route recurrence. The implication is that humans still use movement to regulate attention, even when they do not call it prayer.

Sacred pilgrimages make this regulation explicit and communal. They provide shared routes that can be interpreted, inherited, and corrected. They bind memory to place. They offer a language for change that does not require constant novelty.

The unresolved tension is sharp. When a path is prescribed—by tradition, by authority, by design—does it make you freer to attend, or does it make you obedient without reflection? The orbit still calls. The line still stretches ahead. The test is whether you walk as a conscious point within the pattern, or as a particle pushed by the crowd.

Chapter 5
Nature as First Icon

In 1687, when Isaac Newton published *Philosophiæ Naturalis Principia Mathematica* in London, he did not present his work as a repudiation of nature's mystery, but as its careful reading. The heavens, he argued, moved according to intelligible laws. The same forces that bent an apple's fall governed the sweep of planets. The outcome was revolutionary: the sky became legible. Yet Newton's achievement rested on an older instinct, one shared by shepherds, sailors, monks, and farmers long before him—the instinct to look at the natural world and assume it meant something.

Long before temples were raised, the world itself was the sanctuary.

This chapter turns outward, to what is both most familiar and most easily overlooked. After tracing patterns in geometry, architecture, and ritual, the argument now grounds itself in the first and most enduring source of form: nature itself. Not nature as sentiment or backdrop, but nature as structured presence—spirals that grow without tearing, crystals that assemble without instruction, cycles that repeat with such reliability that whole civilisations have built their time around them.

The claim is not that nature is sacred because it is beautiful. Beauty alone explains nothing. The claim is that nature presents pattern with such consistency, across such scales, that human cultures have repeatedly treated it as a kind of primary text. Before scripture was written, it was watched. Before doctrine was taught, it was counted.

Spirals provide the most immediate entry point. Shells, climbing plants, whirlpools, and hurricanes all curve in ways that are neither random nor rigid. The spiral allows growth without collapse, return without stagnation. It is motion that remembers itself. Early observers did not need calculus to grasp this. They saw that life rarely grows in straight lines. Expansion comes with curvature. Progress involves turning.

Crystals tell a different story. Where spirals speak of growth over time, crystals speak of constraint at origin. Snowflakes, salt lattices, gemstones—these forms arise because atoms lock together according to fixed angles and distances. No guiding hand shapes each facet. Order appears because matter obeys its own limits. Elegance, here, is not imposed. It emerges.

Between these two—spiral and crystal—lies a tension that has always preoccupied human thought. Is order dynamic or fixed? Is meaning discovered through movement or through structure? Nature refuses to choose. It offers both, often in the same system.

Then there are the largest patterns of all: the sky's cycles. Day yielding to night. Moon waxing and waning. Seasons returning with dependable irregularity. These rhythms are not decorative; they are survival information. Early calendars were not philosophical luxuries. They were tools. Plant too early and crops fail. Travel at the wrong tide and ships are lost. Observe accurately, and life becomes possible.

Out of these observations came festivals, fasts, and holy days—time itself shaped by pattern. The year became a circle. History acquired rhythm. Even now, long after artificial light and global supply chains have loosened our dependence, these cycles continue to shape emotion and expectation. Winter still compresses. Spring still releases. Bodies remember what schedules forget.

To speak of nature as an icon is therefore not metaphorical excess. In religious traditions that permit images, icons are not idols but windows—structured surfaces that train attention toward what cannot be seen directly. Nature functions in the same way. It does not answer questions. It provokes them. It presents order without explanation and invites interpretation without coercion.

This invitation has always carried risk. To read nature as text can slide into projection or superstition. Patterns can be over-interpreted. Regularity can be mistaken for intention. This chapter does not defend naïveté. It insists on discipline. Reading requires literacy. Not every curve is a message. Not every cycle is a command.

Yet neither is pattern meaningless simply because it can be explained. Explanation does not exhaust significance. Knowing how a spiral forms does not cancel its power to instruct. Understanding atomic bonding does not empty crystals of wonder. Precision sharpens attention; it does not dull it.

The modern relevance of this argument is acute. At a moment when environmental degradation threatens the very systems that once oriented human life, the question of how we *read* nature becomes ethical. If the world is treated as mute resource, it will be consumed accordingly. If it is treated as patterned order, it demands response. Not reverence in the abstract, but restraint grounded in understanding.

This chapter will move from the small to the vast, from the curl of a shell to the turning of the year. It will examine how simple forms scale, how constraints generate beauty, and how cycles teach patience. It will ask what it means to treat creation as instructive without turning it into moral propaganda or aesthetic escape.

The tension that drives the pages ahead is deliberate. Can we learn to read nature seriously again—neither romanticising it nor reducing it to raw material? Can pattern be recognised as teacher without being worshipped? The world is still speaking. The question is whether we have retained the patience to listen.

Spirals, shells, and the growth of form

In August 2005, as Hurricane Katrina advanced across the Gulf of Mexico toward New Orleans, satellite images released by the National Oceanic and Atmospheric

Administration showed a vast spiral tightening around a darkened eye. The outcome was catastrophic on land, but the form itself was unmistakable: rain bands curved inward with disciplined regularity, each arc feeding the next. The storm did not advance as a wall. It turned. It returned. It gathered strength by circling.

The spiral was not decoration. It was mechanics made visible.

That same curve appears, at radically different scales, in nautilus shells, sunflower heads, pinecones, whirlpools, and galaxies. The recurrence invites a question that refuses to go away: why does growth so often turn instead of stretching straight? Why does expansion prefer curvature?

Begin with the shell. A mollusc grows by adding material to its shell's opening, enlarging the form without altering its shape. The behaviour is accretion at the edge. The metric is proportional increase over time. The implication is continuity: the animal can grow without abandoning its past. The spiral records history as structure.

This is not poetic licence. Measurements of many shells show logarithmic spirals, where the angle remains constant while the distance from the centre increases exponentially. The shell does not "aim" for beauty. It obeys a rule that solves a problem: how to grow larger without rebuilding. Efficiency leaves a curve.

Plants solve a related problem in air rather than calcium. Phyllotaxis—the arrangement of leaves around a stem—

often follows spiral patterns that maximise light exposure and minimise shading. The behaviour is leaf placement at successive angles. The metric is divergence angle, often close to 137.5 degrees. The implication is optimisation under constraint. Each new leaf arrives where it will steal the least light from its predecessors.

This angle is not mystical. It emerges from simple growth rules interacting with space. Yet its consequences are visible and repeatable. Count the spirals on a sunflower head and you will often find consecutive Fibonacci numbers, a fact documented by botanists since the nineteenth century. The numbers matter less than the principle: turning distributes.

Distribution explains storms as well. A hurricane is an engine that moves heat from warm ocean surfaces upward and outward. Rotation arises from the Coriolis effect acting on rising air, but the spiral structure reflects more than physics jargon. It reflects balance. Air must rise, cool, and be replaced. The curve allows continuous intake without collision. The behaviour is sustained circulation. The metric is pressure gradient and wind speed. The implication is persistence. A straight inflow would collapse.

Here the spiral's darker lesson emerges. Curves are not always gentle. The same structure that allows growth can also concentrate force. Hurricanes intensify by feeding on their own rotation. Shells protect; storms destroy. The form is neutral. Its consequences depend on scale and context.

This neutrality is essential. Too often, spirals are romanticised as symbols of harmony without acknowledging their ambivalence. Nature uses the curve because it works, not because it comforts.

The quiet motif threading this discussion is return. Spirals return without repeating exactly. Each circuit revisits a neighbourhood from a different distance. This is growth's compromise. One cannot move outward without passing familiar directions again.

That compromise appears in human experience as well, which is why spirals have become metaphors across cultures. But before reaching metaphor, the mechanics deserve respect.

Consider fluid dynamics. In 1963, Edward Lorenz, working at MIT in Cambridge, Massachusetts, published work showing how simple equations governing convection could produce complex, swirling patterns. The "Lorenz attractor," often drawn as a double spiral, revealed how deterministic systems could behave unpredictably. The behaviour is sensitive dependence on initial conditions. The metric is divergence of trajectories over time. The implication is humility: simple rules can yield forms that resist long-term prediction.

This matters because it corrects a common misunderstanding. The presence of spirals does not imply control. It implies constraint interacting with flow. The curve is a record of forces negotiating.

A counter-example sharpens the boundary. Not all growth produces spirals. Crystals grow by repeating fixed angles, producing straight edges and flat faces. The behaviour is lattice expansion. The metric is atomic bonding angle. The implication is rigidity. Where rules lock, curves vanish. Growth becomes faceted.

The reconciliation is instructive. Spirals appear where systems must grow while remaining open—open to material, energy, or information. When openness is replaced by closure, the curve gives way to the plane. Spirals belong to living and moving systems more than static ones.

Plants illustrate this again at larger scale. Climbing vines spiral because twisting distributes mechanical stress. A straight climb would snap under wind. The behaviour is helical wrapping. The metric is tension distribution along the stem. The implication is resilience through curvature. Turning absorbs shock.

Galaxies push the principle to the cosmic scale. Spiral galaxies, including the Milky Way, organise stars into sweeping arms that rotate around a dense core. The behaviour is orbital motion governed by gravity. The metric is angular momentum distribution. The implication is longevity. The spiral structure persists over billions of years because it channels motion without collision.

At every scale, the spiral solves a version of the same problem: how to move outward without tearing apart, how to accumulate without choking, how to repeat without stagnating.

This is why straight lines feel foreign to growth. Straight lines imply finality. They suggest a beginning and an end with no memory between. Growth, by contrast, carries memory forward. The spiral is memory drawn in space.

Human cultures noticed this long before satellites and equations. Celtic art, Greek ornament, Māori carving, and ancient Near Eastern seals all deploy spirals. These were not idle decorations. They were observations stylised. The curve signalled life's tendency to turn back on itself while continuing.

Yet cultural adoption introduces risk. Once a form becomes symbol, it can be emptied of discipline. Spirals printed on walls do not grow. They merely repeat. The behaviour is static display. The metric is surface repetition without process. The implication is dilution. The symbol survives while the lesson fades.

The lesson is demanding. Growth is rarely linear. Expansion requires revisiting old ground. Development carries echoes. The spiral refuses the fantasy of clean departure.

This refusal has practical implications for how progress is imagined. Economic models often assume straight-line growth. Environmental systems do not comply. Resource extraction accelerates, then feeds back as depletion or collapse. The behaviour is overshoot. The metric is consumption rate exceeding renewal. The implication is that ignoring curved dynamics courts disaster. Nature's spirals warn against linear arrogance.

The spiral also cautions against despair. Because it returns, it allows correction. Systems can revisit earlier states under new conditions. Recovery is possible, but not by reversal. You cannot unwind a spiral without breaking it. You can only continue turning with care.

The unresolved question presses inward. When growth in your own life loops back—when familiar challenges reappear—do you read that as failure, or as the geometry of becoming? The curve is already drawn. The choice lies in how you move along it.

Crystals, snowflakes, and the logic of atoms

In January 1885, in Jericho, Vermont, Wilson Alwyn Bentley photographed a snowflake through a microscope, capturing its hexagonal symmetry in sharp detail for the first time. The outcome was deceptively simple: proof that no two snowflakes were identical, yet all obeyed the same underlying geometry. Bentley would go on to photograph more than five thousand snow crystals over his lifetime. What he revealed was not whimsy, but discipline. Each flake followed strict molecular rules, even as each expressed them differently.

Constraint made the beauty possible.

Snowflakes fall quietly enough to invite sentiment, but their form is brutally precise. Each one assembles as water molecules lock into a lattice dictated by hydrogen bonding angles. The behaviour is molecular alignment during freezing. The metric is sixfold symmetry. The implication is unavoidable: elegance does not require intention. It requires obedience to rule.

Crystals extend this lesson beyond weather into matter itself. Sodium chloride crystallises into cubes. Quartz grows as hexagonal prisms. Diamond arranges carbon atoms into tetrahedral networks that yield unmatched hardness. The behaviour is atomic bonding. The metric is fixed interatomic distance and angle. The implication is stark. What appears as aesthetic choice is the visible trace of invisible law.

This law is not abstract. It can be measured, predicted, and exploited. In 1912, in Cambridge, the physicist William Lawrence Bragg used X-ray diffraction to determine the atomic structure of crystals, demonstrating that internal order could be read from external pattern. The outcome transformed chemistry and materials science. Structure became legible. The macroscopic form of a crystal could now be explained by the microscopic arrangement of atoms.

Explanation did not empty the forms of wonder. It sharpened it.

The recurring claim that understanding "ruins" beauty misunderstands the relationship between rule and result. Crystalline beauty does not survive despite constraint; it exists because of it. Remove the rule and the form collapses into amorphous solid. Glass, for example, lacks long-range atomic order. The behaviour is random molecular arrangement. The metric is absence of repeating lattice. The implication is translucence without geometry. Glass can be elegant, but it does not teach the same lesson. Constraint matters.

Snowflakes demonstrate how strict rules still permit variety. Temperature and humidity influence how ice branches grow. Small fluctuations change growth rates along different axes, producing elaborate dendritic arms or compact plates. The behaviour is differential growth under variable conditions. The metric is branching complexity. The implication is subtle: order does not eliminate diversity. It channels it.

This resolves a false opposition often drawn between freedom and law. In crystalline systems, freedom appears as variation within constraint, not outside it. The rule provides a field. Expression happens inside that field.

The quiet motif threading this discussion is the lattice. A lattice holds without squeezing. It gives shape without micromanaging outcome.

The same logic appears in minerals formed deep underground. Slow cooling allows atoms time to settle into ordered arrays, producing large, clear crystals. Rapid cooling traps disorder, yielding smaller or distorted forms. The behaviour is cooling rate. The metric is crystal size and clarity. The implication is patience. Time allows order to emerge.

Geology thus offers a lesson in tempo. Pressure and duration matter as much as composition. This is not metaphorical excess. It is causal. Granite crystals grow large because magma cools slowly beneath the Earth's surface. Basalt crystals remain small because lava cools quickly at the surface. Constraint plus time produces coherence.

Crystalline order also governs living systems at smaller scales. Bone mineralisation relies on the orderly deposition of hydroxyapatite crystals within collagen matrices. The behaviour is guided crystallisation. The metric is bone density and strength. The implication is function built from order. Life borrows crystalline logic without surrendering flexibility.

A credible counter-example keeps the argument honest. Not all beauty arises from crystalline order. Amorphous systems—clouds, smoke, flames—produce forms that captivate precisely because they lack fixed structure. The behaviour is turbulent flow. The metric is unpredictability. The implication seems to undermine the thesis.

It does not. These forms belong to a different regime. They teach about change rather than stability. The reconciliation is straightforward. Nature uses multiple grammars. Crystals speak of permanence under constraint. Turbulence speaks of flux. Both matter. Confusing them flattens meaning.

Crystals also teach limits. Their perfection is brittle. Apply sufficient force and the lattice fractures along predictable planes. The behaviour is cleavage. The metric is fracture orientation. The implication is vulnerability. Order trades resilience for clarity. This is not a flaw. It is a cost.

Snowflakes dissolve at a touch. Diamonds shatter under sharp impact. Constraint delivers strength and fragility together. This pairing carries philosophical weight. Systems optimised for order often fail catastrophically when pushed beyond design limits. The crystal does not bend. It breaks.

Human technologies that imitate crystalline order inherit this risk. Silicon chips rely on near-perfect crystal lattices. The behaviour is electron flow through ordered structure. The metric is conductivity and error rate. The implication is dependence. Modern civilisation rests on crystalline obedience at atomic scale. When order slips, systems fail.

This dependence deepens the chapter's claim. The logic of atoms is not remote. It underwrites daily life. Phones, computers, medical imaging, solar panels—all rely on crystalline materials behaving exactly as expected. Beauty here is reliability. Elegance is predictability.

The aesthetic response to crystals, however, predates technology. Cultures have long treated crystals as symbols of clarity, purity, or power. Quartz points, jade carvings, and gemstones recur in ritual contexts worldwide. These uses can drift into superstition. That drift does not negate the original intuition: crystals look *lawful*. They invite trust because they appear to obey something larger than chance.

Trust, here, is not belief in healing powers. It is recognition of consistency. A crystal does not surprise. It repeats itself. That repetition reassures.

Snowflakes complicate this reassurance by introducing uniqueness. Bentley's photographs showed that while rules are fixed, outcomes are never exactly duplicated. The behaviour is rule-bound variation. The metric is morphological difference. The implication is humility. Even under strict constraint, the world does not repeat itself mechanically.

This has consequences for how order is imagined. If rules guarantee outcome, control becomes the goal. If rules permit variation, stewardship replaces control. The distinction matters.

The chapter's broader argument—that nature can be read without being sentimental—rests here. Crystals do not *mean* anything in isolation. They *show* something: that matter carries instruction. To read that instruction responsibly is to respect limits. You cannot demand a crystal behave otherwise. You can only learn how it behaves and work with it.

The unresolved question presses outward. When you encounter order that arises from constraint—whether in matter, systems, or lives—do you resent the limits, or do you recognise the elegance they make possible? Snowflakes fall, crystals grow, and atoms align without asking permission. The pattern is already written. The choice is whether to learn from it, or to keep mistaking freedom for the absence of rule.

Patterns in skies, tides, and seasons

In 46 BCE, in Rome, Julius Caesar introduced the Julian calendar, advised by the Alexandrian astronomer Sosigenes. The outcome was administrative and civilisational: the Roman year was reorganised to track the solar cycle more reliably, with leap days added to keep festivals from drifting into the wrong season. This was not a poetic reform. It was an attempt to bring human time into alignment with the sky's arithmetic. A society that could command legions still had to negotiate with sunrise.

The heavens set the timetable.

Long before Caesar, people looked up and counted. They watched the alternation of light and dark, the swelling and shrinking of the Moon, the return of heat and cold, the tides' stubborn obedience to lunar pull. These cycles were not abstract curiosities. They were survival signals. They told farmers when to plant, sailors when to depart, communities when to gather, mourn, feast, or rest. Pattern in the sky became pattern in culture.

Start with the simplest cycle: day and night. The behaviour is obvious: activity rises with daylight and falls with darkness. The metric is work hours and sleep. The implication is biological. Human circadian rhythms—

endogenous clocks synchronised by light—anchor mood, cognition, and hormone release. Even in electric cities, the body remains a creature of sunrise.

Early societies responded by ritualising the boundary between day and night. Evening prayers in Jewish tradition, canonical hours in Christianity, the Islamic calls to prayer—these practices map sacred attention onto the day's turning. The abstraction is devotion. The visible behaviour is repeated pause at certain times. The metric is communal synchronisation. The implication is that time itself can become a shared liturgy.

The Moon adds a second clock, slower and more dramatic. Lunar phases provide a visible calendar, marking roughly twenty-nine and a half days from new moon to new moon. The behaviour is night-sky observation. The metric is phase count. The implication is predictability. A community can coordinate without instruments. The sky supplies a public clock.

This is why so many religious calendars are lunar or lunisolar. The Islamic calendar is purely lunar; Ramadan and the Hajj shift through the seasons because the year is not adjusted to the solar cycle. The behaviour is fasting and pilgrimage timed to moon sighting. The metric is month start determined by observation and, in modern administrations, calculation. The implication is submission to celestial rhythm rather than agricultural convenience.

Jewish timekeeping offers a contrasting solution. The Hebrew calendar is lunisolar, keeping months aligned with lunar cycles but adding leap months to prevent drift from the seasons. The outcome is practical: Passover remains in spring. The behaviour is intercalation—adding a month in certain years. The metric is seasonal alignment. The implication is balance: lunar visibility and solar stability held together.

The Christian liturgical year inherits a version of this complexity. Easter is set by a calculation tied to the spring equinox and the full moon, formalised by the Council of Nicaea in 325 CE. The behaviour is festival timing anchored to both sun and moon. The metric is computed date. The implication is cosmic correspondence: salvation history mapped onto sky rhythm.

These are not mere scheduling quirks. They demonstrate a human impulse that persists: to treat celestial regularities as more than mechanics, as cues for meaning-making. The sky is read as a score, and culture supplies the choreography.

Tides introduce a third rhythm, one that binds sky to sea and turns gravity into calendar. In coastal societies, tide knowledge has always been practical. Fishing, travel, and trade depend on water level and current. The behaviour is departure timed to tide. The metric is tidal height and current speed. The implication is humility: the sea will not negotiate.

Scientific explanation does not diminish this. Newton's gravitational theory, published in 1687, described tidal forces as the result of differential gravitational pull from the Moon and the Sun. The explanation enabled prediction. Tide tables became tools for navigation and commerce. Yet prediction did not replace ritual. Many communities retained traditions tied to tides—blessings of boats, seasonal fishing rites—because the practical and the symbolic were never truly separate.

Seasonal cycles complete the system. The Earth's tilt produces shifts in daylight and temperature across the year. These shifts govern migration, breeding, flowering, harvest. Human societies mirrored them with festivals of sowing and reaping, of light returning, of restraint during scarcity. The behaviour is communal gathering at recurrent points. The metric is annual recurrence. The implication is continuity: life is sustained by return.

The Roman calendar reform under Caesar sought to tame seasonal drift precisely because drift threatened meaning. Festivals tied to harvest lose force when celebrated in winter. The calendar is not neutral; it anchors shared emotion to environmental reality. When it slips, coherence slips.

A credible counter-example complicates any simplistic narrative of "early people worshipped the sky." Not all cultures organised time primarily through celestial observation. Some relied more heavily on ecological cues: rainfall, river flooding, animal migration. Even then, sky and land are rarely separable. The reconciliation lies in recognising that "calendar" is not one thing. It is an agreement between community and environment about what counts as reliable signal.

Ancient Egypt provides a precise example. The annual flooding of the Nile governed agriculture and taxation. Egyptian calendars incorporated celestial observation—particularly the heliacal rising of Sirius—but the practical anchor was the river. The behaviour is agricultural preparation timed to flood. The metric is water level. The implication is grounded cosmology: the heavens matter because they help anticipate the Earth's behaviour.

The same principle holds today, even under modern infrastructure. Many cultural holidays still align with older seasonal cues. Christmas in the Northern Hemisphere sits close to the winter solstice, a time long associated with the return of light. Lunar New Year follows moon cycles. Harvest festivals persist in various forms. The behaviour is family gathering and ritual meal. The metric is annual repetition. The implication is continuity under technological change.

Technology alters how we experience these cycles, but it does not erase them. Artificial light can extend day. Climate control can flatten temperature variation. Digital calendars can detach scheduling from observation. Yet the body continues to respond to daylight, and culture continues to need shared temporal landmarks. Remove festivals and people invent new ones. The need for patterned time is not a relic. It is structural.

This is where the chapter's argument becomes pointed. Patterns in skies and seasons were turned into calendars not merely for efficiency but for coherence. The abstraction is social trust. The visible behaviour is coordinated action—planting together, fasting together, celebrating together. The metric is synchronised participation. The implication is stability. A shared calendar creates shared expectation, and shared expectation reduces conflict.

Calendars also encode power. Who declares the new year? Who decides when festivals occur? Caesar's reform was political as well as astronomical. Later, Pope Gregory XIII's calendar reform in 1582 corrected drift accumulated in the Julian system, but adoption varied by country and confession. The behaviour was delayed acceptance in Protestant regions. The metric was years of divergence. The implication is that timekeeping is never purely technical. It is governance.

The quiet motif threading this discussion is the hinge. Dawn hinges the day. New moon hinges the month. Equinox and solstice hinge the year. Hinge moments invite ritual because they make transition palpable. Humans crave punctuation.

This craving can be exploited. Commercial calendars invent "seasons" of consumption. Social media manufactures cyclical outrage. The behaviour is attention capture timed to predictable triggers. The metric is engagement spikes. The implication is that pattern can be used to coordinate for good or to manipulate. Cycles do not guarantee meaning. They provide scaffolding.

Nature's cycles remain indifferent. They simply repeat. The meaning is made in the human response: whether to attend, to align, to remember, to resist.

The forward test is therefore not astronomical but moral. When you notice a cycle—light fading earlier, the Moon thinning, tides shifting—do you treat it as background noise, or as an invitation to re-anchor attention? A calendar can be a tool or a cage. The hinge is still there. The question is who turns it.

Reading creation as text

In 1623, in Rome, Galileo Galilei published *Il Saggiatore* (*The Assayer*), arguing that the "book of nature" is written in the language of mathematics. The outcome was not merely literary. It reframed authority. Instead of relying on inherited commentary, Galileo insisted that observation and measurement could read reality directly. Nature, in his telling, was not mute matter but a text—one that could be misread, mistranslated, or ignored, but not altered by opinion.

The page does not flatter.

That sentence is the first defence against sentimentality. To treat creation as text is not to treat it as a mirror for our feelings. A real text resists the reader. It has grammar. It has constraints. It corrects you when you impose meanings it cannot sustain. The traditions that speak of nature as a book—whether Christian, Jewish, Islamic, or philosophical—tend, at their best, to stress precisely this: creation teaches by being what it is, not by echoing what we want.

Show the practice. Medieval Christian thinkers used the phrase *liber naturae*, the "book of nature," to describe creation as a secondary source of knowledge alongside scripture. Augustine, writing in North Africa in the late fourth and early fifth centuries, warned against confident claims about the natural world that contradict observed reality, precisely because such errors discredit theological teaching. The behaviour he criticised was dogmatic assertion about nature. The metric was public credibility. The implication was intellectual humility: if creation is God's work, then misreading it is a kind of untruth.

This is not modern "science and religion" debate dressed in antique robes. It is a practical epistemology: nature's regularities can discipline human speech.

Islamic tradition developed a related sensibility. The Qur'an repeatedly refers to *āyāt*—signs—in the natural world: the alternation of night and day, the growth of plants, the diversity of languages. The behaviour encouraged is reflection, *tadabbur*, on observable phenomena. The metric is attentiveness rather than mere assent. The implication is that observation can be a form of devotion, provided it does not collapse into superstition.

Jewish tradition, too, includes a long practice of blessing and gratitude attached to natural phenomena—thunder, rainbows, seasonal change—turning observation into a structured response. The behaviour is spoken blessing at specific encounters. The metric is ritual timing. The implication is restraint: words are not poured everywhere but placed at hinges, where nature forces notice.

These traditions converge on a shared claim: creation is legible. They diverge on what that legibility yields. Some treat it as confirmation of divine order. Others treat it as provocation to humility. The point here is not to adjudicate metaphysics but to clarify the discipline involved in "reading."

Reading requires method. Landscapes, weather, and living things do not come with captions. If they teach, they do so indirectly—through patterns, constraints, and consequences. A drought teaches about dependence. A storm teaches about fragility. A migration teaches about timing. The lessons are not sentimental. They are often unpleasant.

This is where modern romanticism goes wrong. It treats nature as benevolent therapist—always healing, always affirming, always wise in the way we prefer. Real nature is indifferent. It nourishes and it kills. It gives beauty and disease in the same breath. A serious "text" must include this ambivalence.

The quiet motif threading this discussion is the margin. Readers scribble notes in margins, but the margins are not the book. Our interpretations sit beside creation; they are not identical with it. When we confuse margin with text, we drift into projection.

Projection is the credible counter-example that must be faced directly. People have long claimed to read omens in weather, disasters, animal behaviour, and chance encounters. Sometimes such claims harden into fatalism or cruelty. A plague becomes proof of moral failure. A flood becomes justification for exclusion. The behaviour is moral judgement based on natural events. The metric is scapegoating and social punishment. The implication is harm: misreading nature can become an instrument of violence.

The reconciliation is not to abandon the metaphor of nature-as-text, but to refine it. A disciplined reader distinguishes description from interpretation. Rainfall patterns can be measured. Human meaning drawn from them must be held tentatively. A forest fire can be studied in terms of ecology and climate. Moral lessons about preparedness or humility may follow, but they must not pretend to be direct transcripts of divine speech.

Galileo's insistence on mathematics was one attempt to enforce this discipline. It was not the only one. The scientific method, as it developed through figures such as Francis Bacon in the early seventeenth century in England, emphasised systematic observation, repeatability, and scepticism toward wishful inference. The behaviour is controlled testing. The metric is reproducible result. The implication is modest authority: claims must submit to evidence.

Yet science alone cannot decide what nature "means" in the ethical or spiritual sense. It can describe how systems behave, what causes what, and what consequences follow. It cannot tell us how to respond beyond the factual. The temptation is to smuggle response back in under the guise of description, either as romantic meaning or as nihilistic reduction.

Both are errors. Romanticism turns nature into a soothing story. Reductionism turns it into a mute machine. Serious reading refuses both. It treats creation as structured reality that can be known in part and responded to with responsibility.

Responsibility becomes unavoidable when the "text" includes human impact. Climate change, biodiversity loss, and pollution are not external background; they are alterations to the page itself. The behaviour is industrial extraction and emission. The metric is rising atmospheric CO_2 concentrations and measured temperature anomalies. The implication is moral: humans are not merely readers but authors of damage. The metaphor of reading becomes accusation as well as invitation.

This is why sentimentality is not merely intellectually weak; it is ethically dangerous. If nature is treated as endlessly forgiving, limits are ignored. If it is treated as meaningless, exploitation becomes easy. A disciplined reading acknowledges constraint. Systems have thresholds. Species collapse. Soils degrade. Oceans acidify. These are not metaphors. They are consequences.

Traditions that speak of creation as teaching often include precisely this theme of limit. Sabbatical laws in the Hebrew Bible include land rest. Islamic jurisprudence contains concepts of stewardship (*khilāfah*) and prohibition of waste (*isrāf*). Christian monastic traditions often emphasised simplicity and attentiveness to seasons. The behaviour is restraint. The metric is reduced consumption and patterned labour. The implication is that reading creation involves obedience to limits, not merely admiration.

At the level of personal formation, the metaphor can be equally demanding. To read weather as text is to accept that conditions change beyond control. To read landscapes is to accept that place shapes desire and vulnerability. To read living things is to accept dependence: on pollinators, microbes, water cycles, soil organisms. Dependence is not a moral failure. It is a biological fact.

This fact can recalibrate spiritual imagination. Many modern people associate freedom with independence. Nature teaches the opposite. No organism survives alone. The behaviour is mutualism—bees pollinating flowers, fungi partnering with roots. The metric is increased plant nutrient uptake via mycorrhizal networks, documented extensively in ecological research through the twentieth century. The implication is relational ontology: being is being-in-relation.

This is not sentimental. Mutualism includes parasitism. Cooperation and conflict coexist. The lesson is not "everything is harmonious." The lesson is "systems are interwoven, and consequences travel."

The forward test is therefore concrete. When you look at a landscape, a weather system, or a living creature, do you treat it as mere scenery, or as structured reality that can correct you? Do you read for comfort, or for truth? A real text changes the reader. The page remains. The question is whether you will let it resist you, sentence by sentence, season by season.

Chapter 6
Harmony, Rhythm, and the Mathematics of Beauty

In the winter of 1896, in Paris, a small audience gathered to hear Claude Debussy's *Prélude à l'après-midi d'un faune*. The piece unsettled expectations. Its opening flute line hovered rather than declared, drifting between tonal centres with deliberate ambiguity. Some listeners found it intoxicating. Others felt unmoored. The outcome was not merely aesthetic disagreement; it was a confrontation with pattern itself. Debussy had not abandoned structure. He had bent it, delaying resolution, stretching time, playing with proportion. What sounded like freedom was, in fact, an exquisitely controlled mathematics of hesitation.

Beauty, it turned out, could emerge from tension as much as balance.

This chapter turns from visible pattern to audible and embodied pattern—from geometry drawn in space to geometry unfolding in time. Rhythm, harmony, and recurrence do not merely please the ear. They organise expectation, coordinate bodies, and stabilise communities. They do so not by accident, but because they align with numerical relationships that the nervous system registers, whether or not it can name them.

Music is the most obvious case. Sound is vibration, and vibration can be counted. An octave is not a metaphor; it is a doubling of frequency. A fifth is a ratio of three to two.

Long before notation, cultures discovered that certain intervals feel settled while others strain. The feeling is not arbitrary. It reflects how waves interfere, reinforce, or resist one another. Tension and release are physical facts translated into emotion.

Yet music never remains in the abstract. It recruits breath, pulse, and memory. A slow adagio lengthens respiration. A driving rhythm accelerates gait. A repeated refrain invites anticipation. Pattern here is not decoration layered onto experience; it is experience shaped.

The same is true of language when it is spoken with care. Poetry, prayer, and chant harness metre and repetition to coordinate breath and attention. A psalm recited aloud does something to the body. A line of iambic pentameter settles into a walking pace. Rhyme closes loops. Caesurae create pause. The behaviour is synchronised speaking and listening. The metric is breath length and stress pattern. The implication is profound: meaning is carried not only by words, but by timing.

This chapter also widens the frame to include time itself as patterned medium. Calendars, rest days, fasts, and festivals are not merely cultural artefacts. They are deliberate impositions of rhythm on lived time. A weekly day of rest interrupts labour. An annual fast reshapes appetite. A feast punctuates scarcity with abundance. These practices turn chronology into narrative, time into shared form.

Such rhythms do not float above the body. They anchor in it. Pulse, breath, and gait are internal clocks, ticking whether or not we attend to them. Ritual practices across cultures—walking meditation, bowing, prostration, processional movement—use these clocks to bring individuals into synchrony with one another. The body becomes a metronome. Community becomes an ensemble.

What binds these domains together is number in motion. Harmony is ratio heard. Rhythm is counting embodied. Beauty arises not from randomness, but from constraint that invites play. Too much regularity deadens. Too little dissolves coherence. The art lies in proportion.

The modern world, saturated with noise and acceleration, has not escaped these laws. It has merely forgotten them. Algorithms optimise engagement by manipulating rhythmic attention. Workdays fracture natural cycles. Music streams endlessly without silence. The result is not freedom, but fatigue. This chapter argues that rediscovering patterned time is not nostalgic retreat. It is practical wisdom.

The pages that follow will trace how mathematics becomes felt experience: how sound, speech, and shared time organise meaning more powerfully than argument alone. They will examine why some patterns heal while others exhaust, why repetition can steady or suffocate, and how beauty emerges when rhythm honours the limits of bodies and communities alike.

The invitation is not to aestheticise life, but to listen to it more carefully. Pattern is already there—in the beat beneath the sentence you are reading, in the breath that carries it, in the day that frames it. The question is whether you will notice the count, or continue rushing past the music that has been keeping time all along.

Music as number in motion

On a spring Sunday in 1722, in Leipzig, Johann Sebastian Bach presented the first cantata of what would become his annual cycle for the Thomaskirche. The choir entered precisely on cue. The organ established a pulse. Voices stacked in measured intervals. The outcome was devotional and civic at once: worship shaped by arithmetic. Bach was not illustrating theology with sound; he was organising time through ratio. The congregation did not need to know the mathematics to feel its force.

Sound obeys number.

That sentence, spare and unavoidable, sits beneath every musical culture that has ever endured. Music begins as vibration. A string moves. Air oscillates. A membrane flexes. These movements repeat at measurable rates, called frequencies, and when frequencies relate to one another in simple ratios, the ear recognises stability. When the ratios strain, the body notices tension. What follows is not metaphor but mechanics translated into feeling.

Begin with the octave. If one note vibrates at 220 cycles per second and another at 440, the higher sounds like the same note renewed. The behaviour is frequency doubling. The metric is 2:1. The implication is identity across scale. The ear hears sameness even as pitch rises. This is why melodies survive transposition and why voices of different ranges can sing together without confusion.

The fifth, with its 3:2 ratio, introduces contrast without rupture. It feels open, strong, expectant. The fourth, 4:3, leans forward. These sensations are not learned arbitrarily. Hermann von Helmholtz, writing in Heidelberg in 1863 in *On the Sensations of Tone,* demonstrated how simple ratios produce fewer acoustic beats—interference patterns that the ear registers as roughness. The behaviour is wave reinforcement. The metric is beat frequency. The implication is comfort through alignment.

Dissonance, by contrast, arises when frequencies collide more chaotically. Minor seconds and tritones generate rapid beats that the nervous system reads as unrest. The behaviour is interference. The metric is audible fluctuation. The implication is unease that seeks resolution. Music does not invent tension. It reveals it.

Scales organise these intervals into pathways. A scale is not merely a collection of notes; it is a map of expectation. Western diatonic scales privilege certain steps—whole tones and semitones arranged to guide motion toward rest. Other traditions map space differently. Indian rāgas prescribe ascending and descending patterns with characteristic emphasis points. Arabic maqāmāt divide the octave into intervals unfamiliar to Western ears, producing colours of longing or repose that depend on microtonal placement. The behaviour is selective use of pitch space. The metric is interval pattern. The implication is cultural shaping of emotional contour without abandoning numerical ground.

This matters because it dissolves a false opposition. Music is neither purely cultural convention nor brute physics. It is physics filtered through human habit. Ratios constrain what works. Cultures decide how to travel within those constraints.

Rhythm extends the same logic into time. A beat is repetition. A metre is counted repetition with emphasis. Whether in a Bach chorale, a West African drumming ensemble, or a contemporary electronic track, rhythm establishes expectation and then plays with it. Syncopation delays the expected accent. Polyrhythm overlays competing counts. The behaviour is patterned anticipation. The metric is pulse subdivision. The implication is engagement. Attention sharpens when prediction is teased but not broken.

The body responds first. Heart rate adjusts. Breathing entrains. Foot taps. These are not optional reactions. Neuroscientific studies across the late twentieth and early twenty-first centuries have documented entrainment: the tendency of biological rhythms to synchronise with external periodic stimuli. The behaviour is phase alignment. The metric is timing convergence. The implication is that music quite literally organises bodies.

This organisation carries social weight. Marching songs coordinate movement. Lullabies slow respiration. Work songs pace labour. National anthems consolidate collective timing into shared identity. The abstraction is cohesion. The visible behaviour is synchronised action—walking, clapping, singing together. The metric is reduced variance in movement. The implication is trust. Groups that move together tend to cooperate more readily.

A credible counter-example tempers triumphalism. Not all music aims at resolution. Twentieth-century composers such as Arnold Schoenberg, writing in Vienna in the early 1900s, deliberately abandoned tonal centres, constructing twelve-tone rows that avoid traditional consonance. The behaviour is equal distribution of pitch classes. The metric is avoidance of hierarchical interval. The implication appears to contradict the claim that simple ratios anchor musical meaning.

It does not. Schoenberg's music still operates numerically. It replaces consonance-based expectation with serial order. Tension does not disappear; it is reframed. Listeners often report sustained unease because the nervous system receives fewer points of rest. The reconciliation is straightforward. Music can resist comfort, but it cannot escape structure. Even rejection requires rule.

The quiet motif threading this discussion is the return. Music moves away and comes back. Cadences close arcs. Refrains repeat. Even through-composed works rely on memory. Return allows the listener to locate themselves in time. Without it, sound dissolves into noise.

This return has ethical resonance. Tension without release exhausts. Release without tension bores. Beauty arises from proportion between the two. This is not aesthetic taste; it is physiological truth. Cortisol rises under unresolved stress. Dopamine rewards expectation fulfilled. Music teaches this balance viscerally.

Historical traditions noticed early. Pythagorean thinkers in southern Italy in the sixth century BCE linked musical intervals to cosmic order, proposing a *harmonia* that structured both sound and stars. While their cosmology overreached, their intuition about ratio endured. Plato, in fourth-century Athens, warned against musical modes that overstimulated or enervated citizens, recognising that rhythm shapes character. The behaviour he feared was habitual exposure. The metric was moral disposition. The implication was education through sound.

Religious traditions integrated this insight carefully. Gregorian chant limits melodic range and rhythm to avoid distraction, producing an aural environment that steadies breath and attention. The behaviour is restrained melodic motion. The metric is narrow interval set. The implication is contemplation. In contrast, gospel music employs driving rhythm and call-and-response to energise and unify. Different aims, same mechanics.

Modern life complicates the picture. Music now saturates environments continuously. Streaming platforms algorithmically optimise playlists to maintain mood, often flattening dynamic range and rhythmic variation. The behaviour is background sound consumption. The metric is listening hours without active attention. The implication is dulling. When music loses its arcs of tension and release, it becomes sonic wallpaper.

This flattening has consequences. Without silence, release loses meaning. Without rhythmical variety, bodies lose entrainment. Fatigue follows. The mathematics remain, but they are poorly used.

To speak of music as number in motion, then, is not to reduce it to calculation. It is to restore responsibility. Numbers describe what sound does to bodies and groups. They explain why certain patterns heal and others agitate. They clarify why beauty feels earned rather than imposed.

The forward test is immediate. When you listen—truly listen—do you notice where tension gathers and where it dissolves? Do you allow rhythm to organise your breath, or do you consume sound as distraction? Music will keep counting regardless. The question is whether you will move with the numbers, or let them move you without consent.

Pattern in poetry, metre, and breath

On 19 June 1965, in London's Royal Albert Hall, T. S. Eliot's *Four Quartets* were performed in a staged production that treated the poem as spoken music rather than silent text.

The outcome was immediate to anyone in the room: listeners breathed differently. Lines that look spare on the page acquired weight and tempo in the mouth. Pauses became audible architecture. A poem, when voiced, revealed its true medium. It was not ink. It was air.

Breath sets the measure.

That claim sounds mystical until one tries to speak a long sentence aloud without rhythm. Speech collapses into panting. Meaning splinters. The body forces punctuation. Metre, rhyme, and repetition are not ornamental frills laid on top of language; they are devices evolved and refined to coordinate breath and listening, ensuring that thought can travel from one body to another without breaking.

Start with the simplest observation. When you recite a line of verse, you cannot avoid taking breaths at certain intervals. The behaviour is physiological. The metric is lung capacity and phrase length. The implication is structural: lines often reflect the body's comfortable exhalation span. Traditional metres emerged not because poets enjoyed counting syllables for its own sake, but because speech needs manageable units.

English iambic pentameter, the backbone of much Renaissance drama and poetry, typically carries ten syllables per line with alternating stress. Spoken at a natural pace, it fits a single breath for most speakers. The behaviour is one-breath delivery. The metric is syllable count and stress pattern. The implication is endurance. A playwright can sustain long scenes of heightened language because the metre keeps speech breathable.

This is not a claim about perfection. It is about constraint. When Shakespeare stretches a line with extra syllables, the listener feels it. When he breaks the metre, the break registers emotionally. Variation matters because there is a norm to vary from.

Rhyme intensifies this coordination by giving the listener an anchor. In oral cultures, rhyme and alliteration are memory technologies. They reduce cognitive load by making phrases predictable. The behaviour is recall under pressure. The metric is accuracy of transmission. The implication is communal continuity. A hymn remembered by sound can outlast paper.

The *Iliad* and the *Odyssey,* attributed to Homer and performed orally long before they were written down, relied on formulaic phrases and recurring epithets—"swift-footed Achilles," "rosy-fingered dawn"—that fit the metre and aided performance. The behaviour is repeated phrase insertion. The metric is metre compliance. The implication is stability: epic memory carried by patterned speech.

Sacred traditions employ the same mechanisms with greater urgency. The Qur'an was first received and transmitted as recitation, and its rhythmic and phonetic structures support memorisation across vast communities. The behaviour is repeated vocalisation. The metric is accurate oral recall. The implication is preservation. Pattern protects text when paper is scarce and authority depends on fidelity.

In Christianity, the Psalms have been sung and spoken for millennia. Their parallelism and repetition—lines that echo and intensify—shape breath and attention. The behaviour is call-and-response or antiphonal chanting. The metric is synchronised timing between groups. The implication is communal listening. People learn to hold silence for one another's words.

The point is not that all prayer is poetic, or that poetry is always prayer. The point is that recitation harnesses bodily cycles. Breath becomes the clock. Stress patterns become the beat. Meaning rides on rhythm.

Modern cognitive science provides corroboration. Psycholinguistic research has shown that rhythmic regularity improves processing and recall. The brain predicts timing. When timing is stable, comprehension improves. The behaviour is reduced processing difficulty. The metric is faster recognition or improved recall in experiments. The implication is that metre is a cognitive tool, not an archaic constraint.

Yet pattern can also deceive. Here the counter-example is essential. Rhyme can make false statements feel true simply because they are easier to process and remember. Psychologists have described this as the "rhyme-as-reason" effect, documented in the late twentieth and early twenty-first centuries: aphorisms that rhyme are often judged more accurate than equivalent non-rhyming versions. The behaviour is increased perceived truth. The metric is higher agreement ratings. The implication is ethical risk. Pattern can smuggle persuasion.

The reconciliation is not to mistrust rhythm, but to treat it with respect. Pattern enhances transmission. It does not guarantee truth. A well-rhymed lie travels farther than a clumsy truth. Religious traditions have always known this, which is why many insist on disciplined forms—canonised texts, authorised liturgies, trained reciters—to constrain improvisation. Structure protects against manipulation.

The quiet motif threading this discussion is the pause. A pause is not absence; it is shape. In recited poetry and prayer, silence measures what follows. It allows breath to reset and gives the listener a beat to absorb. Remove the pause and words become flood. Flood does not teach. It overwhelms.

This is why metre is not merely about stressing syllables. It is also about spacing. Caesurae—intentional mid-line pauses—create emotional contour. In Anglo-Saxon poetry, such as *Beowulf* (preserved in a manuscript around the year 1000), the line is divided by a caesura, with alliteration tying halves together. The behaviour is segmented delivery. The metric is stress count and alliterative link. The implication is power: the voice sounds hammered, deliberate, suited to communal hearing in halls.

Different languages and traditions use different meters because bodies and phonologies differ, but the underlying logic remains: pattern makes speech fit human physiology and social listening.

Listening is often ignored in discussions of poetry, as if the reader were solitary. Yet recitation assumes community. Even when performed alone, the listener is implied. Rhythm coordinates not only breath but attention between speaker and hearer. It sets expectations. It signals when a thought begins, intensifies, closes.

Consider liturgical repetition: the Kyrie eleison, the Jesus Prayer, the rosary, the mantra. Repeated phrases are sometimes dismissed as mindless. This is misunderstanding. Repetition reduces semantic novelty so that attention can shift from parsing meaning to inhabiting it. The behaviour is sustained repetition without cognitive fatigue. The metric is duration of practice. The implication is depth through steadiness.

This can slide into rote. The same mechanism that enables contemplation can enable disengagement. The reconciliation again returns to the pause: reflection inserted into repetition prevents automation. Many traditions build this in through alternating verses, antiphons, or responsive readings. Structure invites wakefulness.

Modernity has not eliminated these dynamics; it has repurposed them. Advertising slogans use metre and rhyme. Political chants use repetition and stress to produce unity and pressure. Pop music lyrics lean on hook lines that fit breath and beat. The behaviour is crowd chanting. The metric is synchronised vocalisation. The implication is collective identity—for good or ill.

The question is therefore not whether pattern will coordinate breathing and listening. It will. The question is who chooses the pattern and to what end.

A final observation sharpens the argument. Silent reading, now common, can trick us into thinking metre is optional. Yet even silent reading often triggers subvocalisation—tiny movements of speech musculature and imagined sound. The body still participates. The breath still shadows the line. Pattern is felt even when not voiced.

The forward test is practical. When you next encounter a poem or a prayer, do you read it as information or as timed speech? Do you notice where your breath is compelled to pause, where stress pulls emphasis, where repetition steadies or numbs? The pause is waiting. Your lungs will answer honestly.

Time cycles, calendars, and sacred festivals

In September 1793, in Paris, the National Convention introduced the French Republican Calendar, replacing the seven-day week with ten-day "décades" and renaming months to reflect seasonal phenomena. The outcome was immediate and, in the end, unsustainable. Workers complained of fewer rest days. Rural communities resisted the new rhythm. Religious observance, tied to Sundays and saints' days, became harder to police precisely because it was harder to schedule. By 1806, Napoleon abolished the system and restored the Gregorian calendar. A government that could redraw borders could not easily redraw time.

The week fought back.

That episode makes a point more sharply than any philosophical claim: calendars are not neutral charts. They are social architectures. They impose pattern on lived time, and once a pattern is internalised—by bodies, economies, rituals, and memories—it becomes stubborn. Change it and you do not merely adjust schedules; you disturb identities.

Sacred festivals are the most explicit expression of this architecture. Weekly rest days, annual fasts, and recurring feasts are not spontaneous celebrations that happen to repeat. They are deliberate impositions of rhythm on desire, labour, and attention. They turn time into a shared geometry: intervals, cycles, returns, punctuations. Communities do not merely "mark" time. They shape it.

Begin with the week, the smallest durable unit of communal rhythm that is not obviously dictated by astronomy. A day corresponds to the Earth's rotation. A month can be anchored to lunar cycles. A year follows the Sun. The seven-day week is a cultural invention that achieved remarkable persistence. Its endurance suggests that it meets a human need: a stable cycle short enough to organise labour and rest, long enough to carry narrative.

In Jewish tradition, the Sabbath anchors that cycle with theological insistence. The behaviour is cessation from work on the seventh day. The metric is recurring weekly interruption. The implication is radical: time itself can be sanctified. Rest is not merely recovery; it is obedience to a pattern that places limits on production. That limit is social as well as spiritual. It protects the vulnerable by restraining the powerful's appetite for continuous extraction.

Christianity inherited and transformed this rhythm, shifting communal worship to Sunday while retaining the week's geometry. Islam, while not treating Friday as a Sabbath in the same manner, nonetheless structures communal prayer around it, creating a weekly hinge that orients social life. The abstraction is devotion. The visible behaviour is synchronised gathering and altered labour patterns. The metric is weekly repetition. The implication is cohesion: communities become legible to themselves.

Annual fasts and feasts extend this geometry into the year, turning seasons into moral exercises. Ramadan, the month-long fast in Islam, imposes a daily cycle of restraint and release across weeks. The behaviour is abstaining from food and drink during daylight hours. The metric is hours fasted. The implication is discipline distributed across time, not concentrated in a single heroic act. The fast trains patience and empathy through repeated small denials, rather than through dramatic gestures.

Lent functions similarly in many Christian traditions, though its practices vary. The behaviour is chosen abstinence or intensified devotion across roughly forty days. The metric is sustained duration. The implication is preparation: desire is recalibrated in anticipation of feast. The alternation of fasting and feasting is not merely diet. It is a narrative structure: deprivation gives meaning to abundance, and abundance gives reason for deprivation.

Feasts, in turn, are not optional rewards. They are communal affirmations that life includes joy and that joy can be shared without apology. Passover, Easter, Eid al-Fitr, Diwali—each binds story to time. The behaviour is gathered meal, ritual speech, repeated symbols. The metric is annual recurrence. The implication is memory made bodily. History becomes edible.

This is where the phrase "shared geometric frameworks for stories" earns its weight. A calendar is a set of coordinates. It tells a community where they are in the year's narrative. Without that coordinate system, stories become detached from lived experience. They might still be told, but they no longer organise desire and expectation. Sacred festivals refuse this detachment. They insist that meaning must be rehearsed.

The quiet motif threading this discussion is the return. Sacred time returns without being redundant. Each repetition arrives in a new year, with new griefs, new births, new political weather, new private failures. The ritual remains. The participants change. The return allows comparison. This year's fasting feels different from last year's. This year's feast tastes different. The pattern becomes a measuring instrument for interior life.

Modern psychology and sociology offer corroboration without claiming to exhaust meaning. Regular rituals increase predictability, and predictability reduces stress in uncertain environments. The behaviour is adherence to repeated practices. The metric is reduced reported anxiety or increased wellbeing in some studies of religious participation. The implication is stabilisation: ritual time offers a scaffold when personal circumstances shake.

Yet here a credible counter-example must be faced. Pattern can also become coercion. Festivals can be weaponised to enforce conformity, marginalise minorities, or punish those who do not participate. The behaviour is social pressure and exclusion. The metric is penalisation of non-observance. The implication is harm: a calendar can become a cage.

The reconciliation lies in distinguishing shared rhythm from imposed uniformity. Sacred calendars work best when they invite participation rather than compel it. They shape communities through attraction and habit, not through fear. This is not a sentimental claim. It is a historical observation. Communities that rely on coercion often hollow their rituals into performance. Communities that rely on shared meaning sustain them across centuries.

The French Republican Calendar failed not only because it was inconvenient, but because it lacked deep narrative resonance. Renaming months "Brumaire" and "Thermidor" might delight poets, but it did not supply a story capable of competing with inherited sacred time. It offered administration, not meaning. Humans, however, do not live on administration alone.

Calendars also shape economics. Weekly rest days structure markets. Holiday seasons alter consumption. Agricultural festivals coordinate labour. These are measurable behaviours: retail surges, travel peaks, workplace closures. The metric is economic activity shifting predictably around dates. The implication is that sacred time is never purely "spiritual." It reorganises material life.

This reorganisation can be a form of resistance. A Sabbath refuses constant availability. A fast refuses constant consumption. A feast refuses scarcity mentality. In each case, the calendar imposes limits on the market's claim to total time. Even in secular societies, remnants of these rhythms persist as weekends and public holidays. The origins may be forgotten, but the geometry remains.

Digital life has begun to erode these rhythms. Work messages arrive on Sundays. Commerce operates continuously. Streaming and social media remove the need for scheduled entertainment. The behaviour is time flattening—activities spread across all days and hours. The metric is reduced distinction between work and rest. The implication is fatigue. When nothing is set apart, nothing replenishes.

This is why sacred festivals still matter, even to those who do not share the beliefs behind them. They demonstrate that time can be shaped deliberately rather than surrendered to convenience. They show that repetition can protect freedom by limiting choice. Without boundaries, choice becomes compulsion.

The forward test is therefore personal and communal. When your calendar offers no sacred interruptions, what rhythms take their place? Do you allow your weeks and years to be designed by urgent demands and algorithmic prompts, or do you impose patterns that tell a story worth living? The return is coming either way. The question is what it will return you to.

The body as metronome

In 1978, in Amherst, Massachusetts, the psychologist William H. McNeill began writing what he later published as *Keeping Together in Time* (1995), arguing that rhythmic synchrony—marching, dancing, chanting—has repeatedly generated social cohesion across history. His outcome was not a vague celebration of "togetherness" but a specific thesis: when bodies entrain to a shared beat, they become more willing to cooperate, obey, and sacrifice. McNeill traced this from military drill to religious procession. The claim was unsettling because it was measurable in behaviour. Rhythm changes groups.

Your pulse is counting.

Even in stillness, the body keeps time. The heart contracts and releases. The lungs draw air and let it go. The muscles maintain tone and shift weight. Gait, when you walk, settles into a cadence that can be predicted almost as reliably as a metronome. These internal clocks do not merely support life; they shape perception. They determine what feels "too fast," "too slow," "too long," "too abrupt." They are the hidden standards by which external rhythms are judged.

Sacred ritual has always understood this, often more practically than modern people realise. Many rituals do not aim to impress the mind with ideas; they aim to synchronise bodies so that ideas can be carried. Bowing, kneeling, prostration, walking meditation, processional movement—these are technologies of timing. They harness the body's metronomes and align them with communal forms.

Start with breath. Breathing is involuntary, but its rhythm is unusually permeable to influence. A hymn sung slowly extends exhalation. A chant repeated steadily reduces abrupt inhalations. The behaviour is paced respiration. The metric is breath rate and depth. The implication is attentional steadiness. When breath settles, the mind becomes less reactive; when breath fragments, attention scatters.

This is not a spiritual claim masquerading as physiology. Research in psychophysiology has long shown that respiration influences heart-rate variability—a marker of autonomic regulation—though interpretations must remain cautious. The point here is simpler and undeniable: slow, regular vocalisation tends to produce slower, more regular breathing. Ritual uses this without needing a laboratory.

Bowing links breath to posture. In Islam, the sequence of movements in salat—standing, bowing, prostrating, sitting—creates a repeating cycle that coordinates breath, joint motion, and attention. The behaviour is timed bodily transition. The metric is repeated cycles per prayer. The implication is embodied submission: the body rehearses humility not as sentiment but as repeated action.

Japanese Zen practice offers a different but related example. In zazen, posture is held with deliberate stillness, yet walking meditation (*kinhin*) punctuates sitting with slow, measured steps. The behaviour is slow gait under instruction. The metric is step count per breath, often practised with remarkably consistent pacing in monastic settings. The implication is continuity: attention is trained not only in stillness but in motion.

Gait is an internal clock because it is rhythmic and self-sustaining. Walk at a steady pace and you can predict the next footfall without thinking. This frees attention. It also makes it easy for groups to synchronise. Marching works for this reason. Processions work for the same reason, though their intent differs.

Consider Christian processions during Holy Week, or Orthodox Paschal processions that circle the church. The behaviour is collective walking at common pace, often accompanied by chant. The metric is synchronised movement and vocal timing. The implication is shared time made visible. The community becomes a moving pattern.

Hindu and Buddhist circumambulation, too, uses gait to transform devotion into orbit. The body's metronome turns the circle into lived time. Each step is a unit. Each circuit is a paragraph. The pilgrim does not merely think about the sacred centre; they pace it.

The quiet motif threading this discussion is the beat. Not the beat of music alone, but the beat as an agreement: my timing meets yours.

That agreement carries social consequences. Studies in social psychology have repeatedly found that synchronised movement can increase affiliation and prosocial behaviour, though the magnitude and conditions vary. The behaviour is increased helping and perceived closeness after moving in time with others. The metric is cooperative choice in experiments. The implication is that ritual synchrony is not incidental. It is one of the ways communities manufacture trust.

Trust is not an abstraction here. It is visible in what people are willing to do together: sing publicly, kneel, fast, walk long distances, share silence. These behaviours require vulnerability. Synchrony reduces the felt risk of that vulnerability because it signals belonging.

Yet this is where a credible limitation must be faced. Synchrony can also be used to manipulate. Military drills produce obedience. Political rallies use chants to dissolve individual judgement into collective momentum. The behaviour is unified shouting and movement. The metric is reduced dissent signalling. The implication is danger: the same mechanism that binds can also coerce.

The reconciliation lies in attention and consent. Ritual synchrony that aims at contemplation makes room for interior freedom—silence, reflection, voluntary participation. Coercive synchrony aims at erasing interior distance. One can often tell the difference by what happens to silence. Contemplative rituals honour it. Coercive rituals fear it.

The body's metronomes also shape individual experience of time. Anxiety often accelerates internal clocks: heart rate rises, breathing shortens, attention narrows. Rituals that slow breath and regulate posture can counteract this acceleration, not by suppressing emotion but by altering its tempo. Again, this is education, not advice. The structural point is that tempo changes feeling.

This is why many sacred practices repeat. Repetition is not lack of imagination. It is training. A single bow does little. A thousand bows reshape habit. A single chant line is forgettable. A chant repeated daily becomes a groove in attention.

In some traditions, the repetition is explicit. Tibetan Buddhist prostration practices can involve hundreds or thousands of full-body bows. The behaviour is repeated movement with prayer. The metric is count. The implication is perseverance embodied. Counting prevents vagueness. It turns aspiration into measurable commitment.

Even less dramatic practices rely on measurement. Anglican liturgy, with its repeated responses, coordinates communal speech. The behaviour is call-and-response recitation. The metric is shared timing. The implication is unity without uniformity: many voices speak as one without losing their timbre.

Modern life tends to weaken these embodied clocks by fragmenting rhythm. Notifications interrupt. Work schedules shift. Meals become irregular. Sleep becomes negotiable. The behaviour is constant switching. The metric is disrupted routine. The implication is temporal disorientation. People feel "busy" not only because of volume of tasks but because of loss of cadence.

Ritual restores cadence by insisting on patterned movement and pause. It reminds the body that time is not merely something to fill; it is something to inhabit.

The forward test is practical and quietly unsettling. When you participate in a ritual—whether a prayer, a song, a walk, a bow—do you notice what happens to your internal clocks? Does your pulse steady or spike? Does your breath lengthen or fragment? Do you leave more able to choose your attention, or more carried by the crowd? The beat will keep counting. The question is whose time you are living in.

Chapter 7
Pattern in the Mind

In the summer of 1943, the psychologist Wolfgang Köhler described a curious moment from a walk in the countryside. He had been staring idly at a jumble of rocks when, suddenly, a shape resolved itself: a figure standing upright, unmistakably human. Nothing in the scene had changed. The stones lay where they had always been. What changed was Köhler's perception. His mind had stitched fragments into form. He had not discovered something new in the landscape; he had *completed* it. This moment, minor and almost laughable, opens onto one of the most consequential facts about human consciousness: we do not simply receive the world. We assemble it.

Every second, the brain is confronted with far more information than it can process. Light strikes the retina in shifting patterns. Sound arrives as pressure waves. Touch, balance, smell, and internal sensation compete for attention. Out of this flux, the mind produces a stable scene. It draws edges where none exist. It infers continuity across gaps. It prefers regularity over randomness. Pattern is not an ornament added to perception. It is the method by which perception works at all.

This chapter turns inward. Having traced patterns in stone, sound, space, and ritual, we now confront the most intricate pattern-maker of all: the human mind. The question is no longer where pattern appears, but *how we recognise it, why we crave it,* and *when that craving misleads us*. The stakes are higher than curiosity. The same mental habits that allow us to see meaning can also trap us in illusion.

Vision offers the clearest entry point. Look at a cloud long enough and animals emerge. Faces appear in bark, plugs, headlights, and sockets. This is not childish fancy. It is a predictable outcome of a visual system tuned for survival. Missing a predator is costlier than mistaking a shadow. The brain therefore errs on the side of detection. It fills gaps. It extrapolates from partial data. It sees wholes before parts. These tendencies make everyday perception possible. They also explain why we sometimes see what is not there.

Hearing operates similarly. Speech recognition depends on detecting regularities in sound streams, even when syllables blur or noise intrudes. We understand familiar voices on bad phone lines because the brain predicts what should come next and checks incoming sound against expectation. Meaning emerges not from perfect data, but from disciplined guessing. Perception is inference under pressure.

From this emerges a crucial insight that will shape the chapter: attention is not passive. It is selective, predictive, and biased. The mind does not wait for meaning to arrive fully formed. It hunts for it.

Modern neuroscience increasingly describes the brain as a prediction machine. At every level, it generates expectations about what it is about to perceive and updates those expectations when they fail. This framework does not reduce experience to calculation; it explains why experience feels coherent at all. We notice surprises because they violate prediction. We feel boredom when nothing challenges it. We feel anxiety when predictions fail too often.

Prayer, study, and contemplation can be understood through this lens without flattening their depth. Each is a way of tuning attention—training the mind in what patterns to expect, what signals to prioritise, what noise to ignore. A repeated prayer does not inform the mind of new facts. It reshapes expectation. A sacred text read slowly trains the reader to notice certain kinds of connection. Silence teaches the mind to tolerate the absence of immediate pattern and wait.

This does not mean that spiritual practices are merely cognitive tricks. It means they work *with* the architecture of perception rather than against it. They acknowledge that the mind will always search for pattern, and they offer it a disciplined field in which to do so.

But the same architecture carries risk. Pattern-seeking can go astray.

Superstition arises when coincidence is mistaken for causation. A lucky charm appears to "work" because memory highlights hits and forgets misses. A ritual performed before success becomes falsely linked to the outcome. None of this requires foolishness. It requires a normal human mind operating without sufficient correction. The error lies not in seeking meaning, but in failing to test it.

History is filled with such misfires. Omens read in the flight of birds. Astrology derived from planetary cycles. Numerological obsessions that turn arbitrary counts into destiny. These systems often possess internal coherence and aesthetic appeal. They feel convincing because they *are* patterned. The problem is not that they detect patterns, but that they refuse to let patterns be provisional.

This chapter approaches such errors without mockery. To sneer at superstition is to misunderstand its source. The same mental habits that produce science, art, and faith also produce false connections. The difference lies in how those habits are disciplined. Are patterns tested against reality? Are they held with humility? Can they be revised?

The distinction matters because modern life, for all its data, is not less prone to mispatterning. Conspiracy theories flourish where attention is fragmented and trust eroded. Algorithms amplify signals that provoke rather than those that clarify. The mind, flooded with stimuli, grasps at coherence wherever it can find it. When institutions fail to provide shared narratives, individuals invent their own.

Against this backdrop, the chapter proposes a quieter, older response: training perception itself. Not through grand theories, but through modest practices that sharpen attention and slow interpretation. An attentive walk trains the eye to notice without immediately naming. Drawing teaches the hand to follow observation rather than assumption. Chanting stabilises attention through rhythm. None of these practices promise revelation. They promise clarity.

The goal is not to see *more* patterns, but to see *better*. To distinguish signal from projection. To notice when the mind is completing reality too eagerly. To hold meaning lightly enough that it can deepen rather than harden.

There is a moral dimension here. Discernment—ethical, spiritual, relational—depends on accurate perception. Misreading situations, motives, or consequences is often a failure of pattern recognition rather than intention. Learning how the mind constructs meaning is therefore not an abstract exercise. It shapes how one judges, forgives, commits, and waits.

A subtle motif runs beneath the chapter: stitching. The mind stitches fragments into wholes, moments into narratives, actions into identities. The stitching can be careful or careless. It can strengthen or distort. Once seen, it cannot be unseen.

The pages that follow will not offer a system. They will offer a lens. They will examine how the brain makes sense of noise, how attention hunts for order, how error creeps in, and how perception can be trained without becoming rigid. The invitation is simple but demanding: to become aware not only of *what* you see, but *how* you see it—and to ask, patiently, whether the patterns guiding your life deserve the trust you place in them.

How brains stitch meaning from noise

In 1911, the Gestalt psychologist Max Wertheimer was travelling by train between Frankfurt and Würzburg when a flickering light caught his attention at a railway crossing. Two stationary lights, flashed in quick succession, appeared to move. The motion was illusory. Nothing travelled between them. Yet the experience was unmistakable. Wertheimer would later identify this as the *phi phenomenon,* a foundational insight that helped establish Gestalt psychology: the mind does not merely register isolated sensations. It actively organises them into coherent wholes. From that moment, perception could no longer be treated as a passive mirror. It was an act of construction.

That insight still stands. Vision and hearing do not deliver the world to us fully assembled. They offer fragments, delays, ambiguities, and gaps. The brain's task is to make something usable from this partial feed. It does so by detecting regularities, extrapolating continuity, and preferring pattern over randomness. Without these habits, experience would dissolve into static. With them, meaning emerges.

Consider vision. The retina does not record objects. It records contrasts of light. Edges are inferred. Surfaces are guessed. Depth is reconstructed from cues such as shadow, overlap, and relative size. A chair does not arrive in the brain as a chair. It arrives as a shifting pattern of brightness that must be interpreted. This interpretation happens so quickly that it feels immediate, but it is anything but simple.

A classic demonstration comes from the Kanizsa triangle, first described by the Italian psychologist Gaetano Kanizsa in 1955. Three incomplete circles and three angles are arranged so that viewers reliably perceive a bright white triangle floating above them. No triangle is drawn. The brain supplies it. It prefers a coherent shape to a collection of unrelated marks. The implication is profound: perception includes elements that are not present in the stimulus at all.

This same mechanism explains why faces appear everywhere. Two dots and a line are enough. Electrical sockets grin. Cars scowl. The phenomenon, known as pareidolia, reflects the extraordinary sensitivity of facial recognition systems. Neuroscientific studies have identified specialised regions, such as the fusiform face area, that respond preferentially to face-like configurations. These systems evolved under pressure. Failing to detect another human carried social and survival costs. Seeing a face where none exists is the safer error.

Hearing follows similar principles. Sound reaches the ear as overlapping pressure waves. In a crowded room, frequencies collide. Voices mask one another. Yet conversation remains intelligible. The brain separates streams by detecting regularities in pitch, rhythm, and timing. This process, known as auditory scene analysis, relies on prediction. When a voice pauses, the mind anticipates its return. When a familiar melody drops out briefly, the tune continues internally.

This predictive element has been demonstrated experimentally. In 2012, researchers at University College London, led by Karl Friston, showed that the brain actively generates expectations about incoming sensory input and updates them when prediction errors occur. The data do not simply flow upwards from the senses. They are compared against models that flow downwards from prior experience. Perception becomes a negotiation between what is sensed and what is expected.

This helps explain a curious everyday fact: degraded signals are often easier to understand when they are familiar. A scratched recording of a well-known song remains recognisable. A distorted word in one's native language is guessed correctly more often than a clearly spoken word in an unfamiliar tongue. The mind fills gaps using stored patterns. Familiarity reduces uncertainty.

The benefit is efficiency. The cost is bias.

Because the brain prioritises regularity, it can impose structure where none exists. Random sequences are perceived as clustered. Coincidences feel meaningful. This tendency has been measured. In a study by Brugger and Graves in 1997, participants prone to paranormal belief were more likely to detect patterns in random dot displays. The dots were generated by a computer. No hidden structure existed. The difference lay in interpretation, not stimulus.

Yet it would be misleading to frame this simply as error. The same mechanism underpins scientific discovery. Johannes Kepler found laws of planetary motion by detecting regularities in astronomical data. Pattern recognition drives hypothesis formation. The challenge is not to abandon pattern-seeking, but to subject it to discipline.

The brain's method can be summarised simply. It compresses information. Instead of storing every detail, it stores rules. A visual scene is encoded as "edge continues," "surface remains," "object persists." This compression allows rapid response. It also means that perception is always incomplete. The mind wagers that the world will behave as it usually does.

The implications reach beyond sensory processing. Language depends on the same stitching. Words are recognised even when letters are missing or misordered, as long as overall structure remains. Rhythm and metre exploit predictive timing. A poem feels satisfying when expectation and surprise balance precisely. Too much regularity becomes dull. Too much deviation becomes noise.

There is a spiritual resonance here, though it must be handled carefully. Many traditions speak of discernment: the ability to distinguish genuine meaning from illusion. Modern neuroscience does not undermine this idea. It clarifies the terrain on which discernment operates. The mind is not a blank slate awaiting revelation. It is an active interpreter that must learn when to trust its completions and when to suspend them.

A credible limitation must be acknowledged. Pattern recognition is not uniform across individuals or cultures. Experience shapes expectation. What one person sees as meaningful structure, another sees as chaos. Visual illusions vary across populations depending on exposure to carpentered environments, as demonstrated by research into the Müller-Lyer illusion among the San people of southern Africa. Perception is trained by environment.

This variability does not negate shared mechanisms. It shows that pattern detection is plastic. It can be refined. It can also be distorted.

The quiet motif threading this discussion is stitching. The brain stitches fragments into scenes, sounds into voices, moments into narratives. The stitching is usually invisible. Only when it misaligns does it draw attention. Illusions reveal the seams.

The forward question is unavoidable. If perception always involves completion, how does one live responsibly within that fact? When does filling gaps become insight, and when does it become projection? The answer cannot be absolute. It must be practiced. The mind will always stitch. The task is to notice the stitchwork—to know when it holds fast, and when it should be gently unpicked before it hardens into certainty.

Attention, prediction, and the search for signals

In May 1953, at the Massachusetts Institute of Technology in Cambridge, Colin Cherry published research on what he called the "cocktail party problem": the fact that a listener can follow one voice in a crowded room and, in the same moment, miss most of what everyone else is saying. The outcome was not a clever metaphor but a durable scientific question. How does the mind select one signal from the noise? Cherry's experiments revealed a blunt truth about consciousness: experience is not everything that reaches you. It is what your attention *admits*.

Attention is the gate.

That gate is not a moral faculty. It is a biological necessity. The sensory world exceeds the brain's processing capacity by orders of magnitude. To survive, the nervous system must filter. It must hunt for relevance. It must prioritise certain patterns—faces, threats, voices, movement, novelty—and suppress others. The result is not an unbiased view of reality but a workable one.

This filtering begins before conscious choice. In vision, the eyes make rapid saccades, jumping several times per second, sampling rather than scanning continuously. The behaviour is selective fixation. The metric is eye-movement frequency and dwell time. The implication is that the world you "see" is a stitched montage built from brief, chosen glimpses. In hearing, auditory attention separates streams, locking onto one frequency band or temporal pattern while treating the rest as background. The behaviour is stream segregation. The metric is recall of attended speech versus unattended speech. The implication is that perception is already interpretation.

Why do certain stimuli seize attention? Novelty is one answer. So is threat. So is reward. But the deeper answer is prediction. The brain is not merely filtering incoming data; it is comparing that data against expectations, scanning for what fits and, more urgently, what does not.

This predictive view of the brain has been articulated with unusual force in recent decades, particularly through Karl Friston's "free energy principle" (2009) and related predictive processing models. The core claim is that the brain continuously generates predictions about sensory input and updates internal models when prediction errors arise. The behaviour is model updating. The metric is reduced prediction error over time. The implication is radical: attention is, in part, the allocation of resources to whatever most needs explanation.

This makes sense of everyday experience. You notice the sound that is out of place: a sudden crash in a quiet office, a baby's cry cutting through a crowd. Your attention is pulled because the event violates expectation. Conversely, you often miss what is perfectly predictable. The hum of a refrigerator fades from awareness precisely because it conforms to the model.

A five-word sentence belongs here. Prediction drives attention hard.

The mind's search for signals is therefore not neutral. It is guided by what it already believes is likely or important. That guidance can be refined. It can also be corrupted.

Consider expertise. A radiologist learns to detect subtle patterns in medical images that a layperson cannot see. The behaviour is faster and more accurate identification of anomalies. The metric is diagnostic performance. The implication is trained attention: prediction models become more precise, and the brain learns what deviations matter. Similarly, an experienced musician hears intonation errors or rhythmic drift that others miss. The pattern is audible to them because their expectations are sharper.

These examples clarify something that is often misunderstood about "being present." Presence is not merely stillness. It is calibrated prediction. The expert is present to nuances because their mind has learned what signals to expect and how to notice when reality diverges.

Prayer, study, and contemplation can be understood in a similar way without reducing them to psychology. They are attention-training practices. They tune what the mind treats as signal. They shape the internal model of what matters.

Take lectio divina, the Christian practice of slow, repeated reading of scripture, developed in monastic settings over centuries. The behaviour is rereading with pause and reflection. The metric is time spent per passage and the recurrence of certain phrases in memory. The implication is model shaping: the reader's expectations about life, virtue, and meaning are slowly rewired. They become more likely to notice certain patterns—gratitude, temptation, mercy—because the text has trained their predictive framework.

Islamic traditions of Qur'anic recitation and memorisation similarly tune attention. The behaviour is repeated vocalisation with precise rules of pronunciation (*tajwīd*). The metric is accuracy and fluency. The implication is embodied prediction: the mouth learns what comes next, and the mind learns to anticipate meaning through sound. Attention becomes steady because the pattern is stable.

Buddhist meditation practices, though diverse, often include training attention to observe sensations and thoughts without immediate reaction. The behaviour is sustained observation and return when distracted. The metric is reduced wandering over time, often measured experimentally as improved performance on attention tasks. The implication is not blankness but flexibility: prediction becomes less tyrannical. The mind learns that not every internal signal requires action.

This is where spirituality meets cognitive architecture with surprising intimacy. The mind is always predicting. It cannot stop. But it can learn to predict differently, and it can learn not to obey every prediction as if it were command.

A credible limitation must be faced. Predictive models can make perception efficient, but they can also make it rigid.

When expectations harden, attention becomes selective in the worst sense: it ignores disconfirming evidence. This is confirmation bias given a neural mechanism. The behaviour is preferential noticing of supportive cues. The metric is asymmetric recall of confirming versus disconfirming information. The implication is delusion without madness: intelligent people can become trapped in self-sealing narratives simply because their attention has been trained to feed them.

This is not confined to religion. It appears in politics, workplace cultures, and personal relationships. Once a colleague is labelled "unreliable," attention notices every late email and ignores every quiet competence. Once a neighbour is labelled "threat," attention interprets neutral behaviour as suspicious. The predictive machine does not ask for moral permission. It runs.

The reconciliation requires a discipline that many traditions have long cultivated: examination. Not self-accusation, but honest review. Christian examen practices, for example, involve recalling the day's events to notice patterns of gratitude, failure, and grace. The behaviour is daily reflective review. The metric is regularity of practice. The implication is model correction: attention learns to see recurring habits rather than isolated incidents, and the mind becomes less vulnerable to impulsive narratives.

Similarly, scholarly study—whether theological, philosophical, or scientific—trains attention by teaching one to distinguish strong evidence from suggestive coincidence.

The behaviour is slow evaluation of claims. The metric is willingness to revise conclusions. The implication is humility: prediction becomes a hypothesis rather than a verdict.

The quiet motif threading this section is the signal. Signals are not merely found; they are filtered into existence by what the mind is prepared to receive.

Modern technology exploits this. Notifications, recommendation algorithms, and outrage cycles are designed to seize attention by manufacturing prediction errors—surprises, alerts, novelty. The behaviour is compulsive checking. The metric is frequency of device interaction. The implication is attentional capture: the gate is held open by external systems that profit from your scanning.

Against this, contemplative practices can be seen as a counter-technology. They reduce artificial novelty so that genuine signals can be detected: bodily fatigue, moral unease, gratitude, compassion, fear. They do not invent these signals. They create conditions in which the mind can notice them without immediately drowning them in noise.

This is not advice. It is an observation about structure. Attention determines what becomes real in experience. Prediction determines what attention seeks. Practices that tune prediction therefore change lived reality.

The forward test is not abstract. When you enter a room, read a message, meet a stranger, sit in silence—what does your attention hunt for first? Threat? Approval? Opportunity? Fault? Grace? The answer is not merely personality. It is training. Your brain will keep searching for signals. The question is whether you will choose a discipline that makes its search more truthful, or allow the loudest patterns to recruit you without consent.

When pattern seeking goes astray

On 30 October 1938, the Columbia Broadcasting System aired Orson Welles's *The War of the Worlds* from New York City. The programme was a dramatisation, but many listeners interpreted it as live news. The outcome, documented in subsequent public debate and later analysed by Hadley Cantril in *The Invasion from Mars* (1940), was a wave of alarm among portions of the audience: people phoned police, fled homes, and searched the night sky for evidence. The broadcast did not create fear from nothing. It offered suggestive fragments—urgent tone, interruptions, apparent eyewitness reports—and listeners stitched them into a coherent pattern: invasion.

Meaning assembled itself fast.

This is the central danger of the pattern-making mind. The habit that usually saves us—detecting signals quickly, filling gaps, committing to an interpretation before the window closes—can become a liability when the signal is ambiguous, the context is unstable, or the desire for certainty is high. Pattern seeking goes astray not because people become irrational creatures overnight, but because they remain human under pressure.

Superstition, in its mild forms, is often the domestic expression of this pressure. The lucky charm tucked into a pocket before a presentation. The "special" pen used for exams. The athlete's ritual that must be performed *exactly* or else. These habits are sometimes laughed at, yet they persist among highly educated people precisely because they serve a function. They offer a sense of control when outcomes are uncertain.

The behaviour is measurable. A person repeats a ritual before a high-stakes event. The metric is the ritual's consistency, not its causal power. The implication is emotional regulation: the ritual reduces anxiety by creating a familiar sequence. It is a psychological tool disguised as causal belief.

This disguise matters. Sometimes the person knows, on reflection, that the charm does nothing. Sometimes they half-believe. Sometimes they fully believe. The spectrum is wide. The common thread is the mind's hunger to connect action with outcome.

A landmark demonstration of this hunger came from B. F. Skinner in 1948, at Harvard University, with what he called "superstition in the pigeon." He delivered food to pigeons at fixed intervals regardless of what the birds were doing. The outcome was striking: pigeons began repeating whatever action happened to precede food delivery—

turning in circles, bobbing heads—as if their behaviour caused the reward. Skinner argued that accidental reinforcement could generate ritualistic behaviour. The behaviour was repetitive movement. The metric was increased frequency of a specific action. The implication was sobering: a brain can infer causation from coincidence if the timing is persuasive.

Humans are not pigeons. Humans have language, culture, and self-reflection. Yet humans share the same learning machinery that links sequences together. When a good outcome follows an action, even accidentally, the mind takes note. If the stakes are high, it takes note aggressively.

This helps explain the "hot hand" belief in sports and the gambler's fallacy in casinos. A player believes they are "on a streak." A gambler believes a roulette wheel is "due" for red after a run of black. These beliefs persist despite mathematical explanations because they satisfy a deeper cognitive craving: a sense that randomness is readable.

The visible behaviour is altered decision-making—placing larger bets, taking riskier shots, clinging to routines. The metric is deviation from baseline strategy. The implication is cost. In gambling, the cost can be financial. In professional settings, it can be reputational. In relationships, it can be relational stability sacrificed to imagined signs.

Coincidence itself becomes a kind of intoxicant. Two people think of each other and then meet unexpectedly. A song plays at the "right moment." A particular number appears repeatedly. The mind asks: what are the odds? The honest answer is often: not as low as it feels.

Probability is notorious for humiliating intuition. Large populations generate surprising coincidences routinely. In a city of millions, improbable matches become ordinary. Yet the emotional experience of coincidence is not purely statistical. It carries a felt significance that mathematics does not erase. To dismiss that feeling is to misunderstand it.

The feeling is a signal, but not necessarily about the external world. It may be a signal about internal salience—what matters to you, what you fear, what you hope. The coincidence becomes meaningful because it intersects with your existing concerns.

Here a credible limitation must be stated plainly. Not every interpretation of "signs" is superstition. Many religious traditions speak of providence, guidance, or calling, and their adherents interpret events through that lens. It would be intellectually lazy to classify all such interpretations as mere pattern error. The difference lies in how the interpretation is held and tested.

A disciplined religious sensibility tends to include safeguards: counsel from others, consistency with moral teaching, willingness to wait, and refusal to treat private signs as public proof. The behaviour is deliberation rather than impulse. The metric is time taken before acting and the involvement of community. The implication is restraint: meaning is sought without being forced.

Undisciplined pattern seeking, by contrast, tends to intensify isolation. The person relies on private interpretation alone. They become less corrigible. Their model of reality becomes self-sealing. The behaviour is withdrawal from corrective feedback. The metric is reduced consultation and increased certainty. The implication is escalation: superstition hardens into paranoia or grandiosity.

This distinction matters because pattern seeking goes astray most dangerously when it becomes immune to correction.

The quiet motif threading this discussion is the charm. Not only the physical charm in a pocket, but the charm of explanation—how seductive it feels to possess a pattern that "makes sense of everything."

Conspiracy theories exemplify this seduction. They offer comprehensive narratives in which nothing is accidental. They convert complexity into intention. They turn random failure into purposeful design. The behaviour is compulsive information seeking and selective sharing. The metric is time spent consuming and circulating reinforcing content. The implication is distrust institutionalised: relationships strain, communities fracture, and real evidence becomes less persuasive than narrative coherence.

This is not a modern novelty. The human mind has always produced conspiratorial interpretations, especially under social stress. What has changed is scale and speed. Digital platforms allow pattern narratives to propagate rapidly, often detached from the slow constraints of verification. The same cognitive machinery Skinner observed in pigeons—timing, reinforcement, repetition—can be exploited by systems that reward outrage and certainty.

Yet the goal here is not to moralise. It is to understand.

Understanding begins with admitting that pattern error is not a pathology reserved for "other people." It is a feature of ordinary cognition. Even scientists, trained in scepticism, fall prey to confirmation bias. Even statisticians feel the tug of lucky numbers. The difference is not purity. The difference is discipline.

Discipline does not mean cynicism. It means learning to ask better questions of one's own interpretations. What else could explain this? What evidence would change my mind? Am I confusing comfort with truth? Do I want this to be a sign because I fear making a decision without one?

These questions are not weapons against belief. They are tools for discernment. They protect faith from becoming magical thinking and protect scepticism from becoming contempt.

The forward test is personal and sharp. When you feel the pull of a pattern—an omen, a coincidence, a "sure sign"—do you treat that pull as a command, or as an invitation to slow down and examine what you are really seeking? The mind will always reach for the charm. The question is whether you will let the charm become your compass.

Training perception as a spiritual practice

In 1907, at the Benedictine abbey of Maria Laach in Germany, Abbot Ildefons Herwegen began reforming monastic education with an unusual emphasis: *Schulung der Wahrnehmung*—the training of perception. The outcome was not aesthetic refinement for its own sake, but moral clarity. Monks were instructed to walk the cloisters attentively, chant with exact rhythm, copy manuscripts by hand, and observe the play of light across stone at fixed hours of the day. Herwegen believed that if perception was careless, conscience would follow.

He was not inventing a novelty. He was restoring an old conviction: how you *see* determines how you judge.

This conviction runs quietly through religious history. Long before "mindfulness" entered secular vocabulary, traditions understood that perception could be disciplined, dulled, or refined. They treated attention not as a personality trait but as a moral instrument. If attention wandered, desire followed it. If attention sharpened, discernment improved.

The premise is simple. Perception is trainable. The consequences are not.

Training perception does not mean seeking visions or heightened states. It means learning to notice what is already there without distortion. The behaviour is sustained, deliberate noticing. The metric is consistency over time, not intensity. The implication is reliability: a mind that sees more accurately is less easily deceived, by itself or by others.

One of the most elementary forms of this training is the attentive walk. In many traditions, walking is slowed and ritualised not to romanticise nature but to stabilise awareness. Zen kinhin, the walking meditation formalised in Japanese monasteries by the thirteenth century, breaks movement into deliberate phases. Step, shift, breathe. The behaviour is synchronised gait and breath. The metric is pace—often measured in steps per breath. The implication is alignment: attention settles because the body supplies a steady pattern.

A similar logic governs the Christian practice of the pilgrimage walk, whether on the Camino de Santiago or in local prayer circuits around churches and fields. The route is fixed. The pace is moderate. The repetition is intentional. Over time, external scenery becomes less distracting and internal chatter loses urgency. The mind stops hunting novelty and begins registering continuity.

This is not escapism. It is calibration.

Drawing offers another route. Medieval novices were trained to copy geometric diagrams, illuminated letters, and architectural plans before they were permitted to interpret texts. The behaviour is careful replication. The metric is accuracy of proportion. The implication is humility: the hand must submit to the form. Modern cognitive research supports the value of this submission. Studies on observational drawing show increased activation in visual processing areas and improved discrimination of detail. The act of drawing forces the mind to abandon shorthand labels—*leaf, face, window*—and attend to angles, negative space, and relation.

You cannot draw what you refuse to see.

Chanting operates on a different axis but serves the same end. Whether Gregorian chant codified in the early medieval West, Vedic chanting preserved through oral transmission in India, or Qur'anic recitation governed by *tajwīd,* the structure is patterned sound repeated precisely. The behaviour is vocal repetition within strict constraints. The metric is fidelity to pitch, timing, and pronunciation. The implication is entrainment: breath, hearing, and memory align.

Neuroscientific studies of chanting and repetitive vocalisation indicate reductions in autonomic arousal and increased coherence in breathing patterns. The effect is not mystical in itself. It is regulatory. A regulated nervous system perceives more accurately because it is less hijacked by urgency.

Here lies an important clarification. Training perception is not about *adding* meaning. It is about removing noise.

A credible counter-example must be acknowledged. Training practices can become mechanical. Ritual can hollow out. A person may walk, chant, or draw with impeccable form while remaining morally obtuse. History offers no shortage of disciplined practitioners who failed spectacularly in compassion or judgement. The limitation is real.

The reconciliation lies in intention and reflection. Practices that train perception must remain connected to interpretive accountability. The behaviour is periodic review—asking what one has noticed and how it has shaped decisions. The metric is coherence between perception and action. The implication is integration: seeing clearly must inform living rightly.

Without this loop, training risks becoming aesthetic self-soothing.

The quiet motif threading this discussion is sharpening. Not sharpening into aggression, but into focus. A dull blade tears. A sharp blade cuts cleanly. A dull perception blurs moral edges. A sharpened one distinguishes.

Consider the Ignatian tradition of spiritual exercises developed by Ignatius of Loyola in the sixteenth century. The exercises combine imaginative contemplation with disciplined review. One observes inner movements—consolation, desolation, resistance—not to indulge them but to name them. The behaviour is attentive noticing of affect. The metric is ability to distinguish patterns over time. The implication is discernment: decisions are evaluated by their fruits, not their emotional intensity.

This is pattern recognition applied inwardly.

Even silence itself can be a training ground. In Quaker meetings, silence is not emptiness but shared attentiveness. Participants sit without agenda, listening for promptings that arise. The behaviour is collective stillness. The metric is restraint—how rarely one speaks. The implication is communal filtering: not every thought deserves expression. Meaning is tested by patience.

Modern life militates against such patience. Continuous stimulation rewards rapid pattern completion. Headlines invite instant interpretation. Social media platforms reward premature certainty. Under these conditions, perception degrades not because people lack intelligence, but because they lack intervals.

Intervals matter.

Training perception reintroduces intervals—between stimulus and response, between noticing and concluding. These gaps are not wasted time. They are the space in which discernment occurs.

This matters ethically. Many moral failures are failures of perception before they are failures of will. People harm others because they misread situations, oversimplify motives, or ignore inconvenient details. Training perception does not guarantee virtue, but it reduces avoidable error.

The forward tension is unavoidable. You already train your perception, whether intentionally or not. Your habits—what you read, how you move, what you repeat—are shaping what you notice and what you miss. The question is not whether your perception is being trained, but by whom, and toward what end.

Chapter 8
False Orders and True

On the morning of 15 April 2013, as smoke rose near the finish line of the Boston Marathon, millions of people reached for their phones. Within hours, online forums were circulating names, photographs, arrows, circles. Patterns bloomed faster than facts. By the evening, an innocent student had been misidentified, his image shared thousands of times as a suspect. The outcome was not merely error but harm. What failed that day was not intelligence. It was discernment under pressure.

Human beings cannot tolerate disorder for long. When shock breaks the surface of events, the mind rushes to impose shape. It links fragments. It draws lines. It prefers a bad explanation to none at all. This impulse is ancient, but the speed and scale with which it now operates are unprecedented. Pattern no longer emerges slowly through shared reflection; it arrives prepackaged, trending, algorithmically amplified.

This chapter turns toward the shadow side of pattern. Not to discredit the search for order, but to rescue it from distortion.

The danger is subtle. The same cognitive habits that allow us to learn, to pray, to read meaning in ritual and nature can also warp perception. Confirmation bias selects what fits an expectation and quietly discards what does not. Selective memory polishes supportive episodes and lets inconvenient ones fade. Coincidence acquires intention. The behaviour is familiar: we notice the hit, forget the miss. The metric is asymmetry of recall. The implication is a story that feels coherent and is quietly false.

Spiritual lives are not immune to this process. In fact, they may be especially vulnerable. When people expect the universe to speak, they may stop asking whether they are listening well. Signs multiply. Interpretations harden. Prayer becomes a decoding exercise rather than a posture of attention. Faith, once meant to orient a person toward truth and charity, can become a private pattern engine, endlessly confirming itself.

The book has so far argued that pattern is real, formative, and often trustworthy. This chapter insists on a harder truth: not every pattern deserves allegiance. Some are impostors, convincing in form and destructive in effect.

Conspiracy thinking offers a stark example. It is tempting to treat conspiracies as failures of education or intelligence. The evidence does not support that comfort. Elaborate conspiracy systems often display impressive internal consistency. They recruit data points, anticipate objections, and reward the initiate with a sense of privileged sight. Their power lies elsewhere. They offer a hidden order in a world that feels increasingly ungovernable.

Seen from this angle, conspiracy is broken theology. It promises omniscience without humility and providence without love. It replaces trust with suspicion and community with secret knowledge. The behaviour is relentless interpretation. The metric is narrative closure: every event is already explained. The implication is isolation. Once the pattern explains everything, nothing and no one can correct it.

This matters because conspiracy thinking damages the social fabric in precisely the places where shared meaning is most needed. It corrodes trust in institutions, yes, but more deeply it corrodes trust in neighbours. It teaches the mind to read malice into ambiguity and intention into accident. In doing so, it trains a habit of fear.

The modern pattern priesthood does not wear robes. It writes code.

Recommendation engines and social feeds now surface patterns in behaviour with extraordinary precision. They predict what you will watch, buy, read, and believe next. Their power is statistical, not moral. The behaviour they optimise is engagement. The metric is time and interaction. The implication is exposure shaping: attention is guided not toward what is most true or wise, but toward what is most likely to hold you.

These systems are not conspiratorial in the classic sense. They do not require hidden villains. They require only incentives. Yet their effect on perception can be profound. They present a curated world in which certain patterns appear ubiquitous and others vanish. Over time, the feed becomes a teacher. It trains expectation.

To call these systems "new priests of pattern" is not melodrama. Priests have always mediated meaning, selecting texts, framing interpretation, shaping rhythm and repetition. Algorithms now perform similar work without accountability to truth, goodness, or beauty. They are powerful because they are invisible.

This chapter does not argue for withdrawal from technology. It argues for discernment. Discernment asks different questions from consumption. Not "does this resonate?" but "what is this training me to expect?" Not "does this confirm what I feel?" but "what does this exclude?" The behaviour is reflective interruption. The metric is diversity of exposure and tolerance for ambiguity. The implication is freedom: the mind regains the ability to choose which patterns to trust.

Yet even the most disciplined discernment meets a limit. There are experiences that resist mapping altogether. Grief does not arrive with geometry. Sudden grace does not submit to ratio. Wonder breaks patterns even as it reveals meaning. To insist on total explanation here is to commit a different error: to confuse comprehension with control.

Humility before what cannot be mapped is not a retreat from reason. It is fidelity to reality. Some truths are known by endurance rather than analysis. Some meanings are carried rather than solved. The behaviour is silence or waiting. The metric is patience. The implication is reverence.

This return to limits completes the arc. False orders tempt us with mastery. True order invites us into relationship. The difference is not subtle in its consequences. One breeds certainty without compassion. The other sustains trust without illusion.

This chapter will move through these tensions without caricature. It will examine how cognitive bias bends spiritual life, how conspiracy mimics faith while hollowing it out, how algorithms quietly catechise attention, and why restraint before mystery is not failure but wisdom. The aim is not to leave the reader suspicious of pattern, but to leave them capable of honouring it rightly.

The page turns on a simple, difficult question. When an order presents itself—intellectually neat, emotionally satisfying, morally charged—how will you test whether it is false or true?

Superstition, coincidence, and confirmation bias

On 13 October 1917, an estimated crowd of tens of thousands gathered in Fátima, Portugal, after months of reports that the Virgin Mary would appear. Rain fell steadily. According to newspaper accounts from *O Século* and *Diário de Notícias* published days later, many witnesses claimed to see the sun "dance" in the sky, spinning and plunging toward the earth before returning to its place. The outcome was enduring devotion, pilgrimage, and a permanent place for the event in Catholic memory. Others present reported seeing nothing unusual at all.

The divergence is not incidental. It is the story.

Here, in a single field, lies the problem this chapter confronts. When expectation is high, emotion intense, and meaning already anticipated, perception bends. Some see a sign. Others see weather and light. Memory consolidates selectively. Testimony hardens. A pattern forms—not simply in the sky, but in the mind.

Superstition begins at this seam between uncertainty and desire. It is not foolishness. It is cognition under pressure.

Human beings are exquisitely sensitive to correlation. We notice when two events occur together and infer connection. This ability is indispensable. Without it, learning collapses. Yet the same mechanism misfires easily. When correlation appears without causation, the mind often refuses to let it go.

The behaviour is straightforward. An outcome follows an action. The metric is temporal proximity. The implication is meaning: *I did this, then that happened*. Once the association forms, repetition strengthens it. Misses fade. Hits remain vivid.

This is selective memory at work. Psychologists have documented this bias extensively, from early work by Frederic Bartlett on memory reconstruction in the 1930s to modern studies of belief perseverance. Memory does not record neutrally. It edits. It favours coherence over completeness.

Confirmation bias sharpens the effect. Once a belief forms—*this charm protects me, this prayer guarantees success, this sign predicts disaster*—attention begins searching for reinforcement. Supporting instances are noticed and rehearsed. Contradictions are discounted as exceptions or tests. The behaviour is uneven evidence gathering. The metric is disproportionate recall of confirming cases. The implication is self-reinforcement: belief becomes harder to dislodge precisely because it is partially right some of the time.

This matters profoundly in spiritual life, where meaning is not only sought but cherished.

Consider the everyday superstition that follows many religious people without ever becoming official doctrine. A prayer is answered, vividly and unexpectedly. Gratitude floods in. A similar prayer, offered later, seems to go unanswered. The memory of the first remains bright; the second dulls. Over time, the believer develops a private theory of divine behaviour—what "works," what provokes favour, what displeases. The behaviour is adjustment of practice based on perceived outcomes. The metric is persistence of the theory despite mixed evidence. The implication is distortion: faith quietly slides into technique.

This slide is rarely acknowledged because it feels reverent. Yet its structure mirrors the gambler's fallacy or the athlete's lucky socks. The content is sacred. The mechanism is the same.

A counter-example must be honoured. Religious traditions have long wrestled with this danger and built correctives. The Book of Job stands as an explicit refusal of simple moral causation. The righteous suffer. The wicked prosper. Any theology that insists otherwise is rebuked. The behaviour encouraged is endurance without explanation. The metric is refusal to infer guilt from misfortune. The implication is maturity: faith that resists superstition by confronting randomness without panic.

Similarly, in rabbinic Judaism, warnings against divination and omen-reading are paired with robust interpretive traditions. The believer is trained to seek meaning in law and ethical obligation, not in chance occurrences. The behaviour is disciplined interpretation of texts rather than events. The metric is stability of practice independent of circumstance. The implication is resilience.

These correctives show that superstition is not belief taken too seriously, but belief taken too privately. It bypasses communal testing.

Coincidence complicates the picture further. Humans underestimate how often unlikely events occur. In large populations and long timelines, improbabilities are guaranteed. Two people share a birthday. A dream resembles a later event. A thought coincides with a phone call. The emotional impact of these moments is disproportionate to their statistical weight.

This disproportion is not irrational. It signals relevance. The coincidence stands out because it touches something already meaningful. The mind responds, not to probability, but to salience.

The visible behaviour is emotional tagging—heightened arousal and memory encoding. The metric is vivid recall long after similar non-events have vanished. The implication is narrative temptation: coincidence feels like a message because it arrives charged.

Problems arise when coincidence is treated as instruction.

Spiritual discernment becomes warped when private coincidences are granted public authority or moral urgency. A person makes a life-altering decision because of a song lyric overheard, a random verse encountered, a sequence of numbers noticed repeatedly. The behaviour is decisive action based on minimal data. The metric is absence of corroboration. The implication is fragility: when outcomes disappoint, trust collapses or blame turns inward.

This is not to deny that moments of insight can arrive unexpectedly. Many traditions allow for sudden illumination. The distinction lies in posture. Illumination invites testing. Superstition demands obedience.

The quiet motif threading this discussion is the match. When events "match" expectations, satisfaction follows.

The mind relaxes. When they do not, tension arises. Confirmation bias is the mind's effort to preserve the match at all costs.

Modern environments intensify this effort. Digital platforms reward confirmation relentlessly. Search results, feeds, and recommendations adapt to past behaviour, offering more of what has already engaged attention. The behaviour is recursive exposure. The metric is narrowing diversity of content. The implication is accelerated bias: belief systems harden quickly because they are rarely contradicted.

Spiritual communities are not exempt. Online sermons, inspirational clips, and devotional content can be consumed in isolation, detached from accountability. The believer curates a private canon that reflects personal preferences rather than shared wisdom. Superstition thrives in this solitude because nothing interrupts it.

The reconciliation does not require cynicism. It requires habits that expose belief to friction. Historical traditions placed spiritual interpretation under rule, council, and time. They insisted that meaning unfold slowly and publicly. The behaviour is consultation and delay. The metric is patience before conclusion. The implication is trustworthiness: interpretations that survive scrutiny are less likely to be spurious.

At its worst, unchecked superstition turns faith inward, making the self the centre of cosmic attention. At its best, disciplined belief directs attention outward, toward responsibility, compassion, and truth.

The forward test is demanding. When an event feels charged—when coincidence glows, when a pattern seems obvious—will you ask not only *what does this mean to me?* but *how might I be wrong?* The answer to that question determines whether pattern becomes wisdom or illusion.

Conspiracy as broken theology

In August 1921, in New York City, *The Dearborn Independent*—a newspaper owned by Henry Ford—began publishing a series of articles alleging a vast Jewish conspiracy behind modern finance, politics, and culture. The outcome was not confined to one town or one readership. The series drew heavily on *The Protocols of the Elders of Zion,* a forged text first published in the Russian Empire in 1903, and helped launder its claims into American respectability. From there, the pattern travelled. It did not persuade by evidence. It persuaded by architecture: an invisible enemy, a hidden plan, a world finally explained.

Certainty can feel like mercy.

Conspiracy thinking does not begin with stupidity. It begins with a human appetite for hidden order—an appetite that, in healthier forms, fuels scientific discovery, historical investigation, and spiritual longing. The conspiracist's claim is not simply that something bad happened, but that nothing significant is accidental, that chaos is a costume worn by intention. This is why conspiracy can feel comforting. If events are engineered, then the universe is legible. If there is a plot, there is a plotter. If there is a plotter, there might be a way to fight back.

This is where the theological fracture appears. Many religious traditions hold that reality has meaning, that events are not merely random collisions. Conspiracy takes that impulse and corrupts it. It offers providence without love, judgement without mercy, revelation without humility. It mimics the shape of faith while gutting its moral centre.

The first move is interpretive. Conspiracy thinking trains attention to treat ambiguity as evidence. A missing piece is not a gap; it is proof of concealment. A contradiction is not a problem; it is a decoy. The behaviour is relentless re-interpretation. The metric is the theory's ability to absorb counterevidence without changing. The implication is insulation: the story becomes unfalsifiable, which is another way of saying it becomes untethered from reality.

The second move is social. Conspiracy thinking turns knowledge into a badge. It divides the world into the awakened and the deceived, the initiated and the sheep. The believer gains an identity that feels heroic: the one who sees through appearances. This identity is intoxicating because it transforms anxiety into purpose. The behaviour is public signalling of secret knowledge. The metric is increased sharing, recruiting, and boundary policing. The implication is fragmentation: communities split into those who "get it" and those who are suspected.

The third move is moral. Once a group is defined as secretly responsible for everything wrong, ordinary restraint collapses. Suspicion becomes virtue. Cruelty becomes vigilance. The target is no longer an individual whose actions can be judged; it is a category whose existence is treated as threat.

History supplies brutal clarity here. *The Protocols of the Elders of Zion* is a textbook example of conspiracy as pseudo-theology. The text claimed to reveal a hidden council orchestrating world events. Its power lay in its explanatory totality: banking crises, political unrest, cultural change—all reduced to a single malevolent will. The behaviour it produced was not mere belief but action: discrimination, violence, scapegoating. The metric was social permission to blame a whole people. The implication was catastrophe, not because the forged text was uniquely persuasive, but because it offered a pattern that answered fear.

A second named example shows how conspiracy can flourish even without ancient prejudice. On 9 February 1950, in Wheeling, West Virginia, Senator Joseph McCarthy claimed to possess a list of communists in the U.S. State Department. The details of the "list" shifted, but the structure held: infiltration, betrayal, hidden networks, compromised institutions. The outcome was measurable: investigations, blacklists, careers destroyed, and a culture of suspicion that corroded civil trust. Here, too, the theory's strength lay less in proof than in posture. Doubt was reframed as naïveté. Disagreement became evidence of complicity.

In both cases, conspiracy offered a grim comfort. It reduced complex systems—economies, governments, cultures—into a single narrative of control. It promised that the world could be read like a code.

This is why conspiracy is best understood as broken theology rather than merely bad politics. Theology, at its serious best, holds two truths in tension: that reality is meaningful, and that human beings see only in part. Conspiracy holds the first truth and murders the second. It refuses partial sight. It cannot tolerate mystery. It is allergic to "we do not know yet."

That allergy has psychological roots. Uncertainty feels like vulnerability. It forces the ego to admit dependence and limitation. Conspiracy replaces vulnerability with mastery: *I know what is really happening.* The believer's anxiety drops not because the world has become safer, but because it has become narratively contained.

Containment is the bait. Community is the casualty.

Trust is never an abstract virtue; it is a lived practice. It is the willingness to grant others the benefit of interpretive charity. It is the choice not to treat every mistake as malice. Conspiracy thinking trains the opposite. It teaches the mind to interpret incompetence as intention, accident as strategy, and complexity as cover. The behaviour is attribution of malicious motive to ambiguous events. The metric is reduced tolerance for uncertainty and error. The implication is social brittleness: relationships become fragile because every disagreement is potentially betrayal.

This brittleness spreads through institutions. Journalism becomes propaganda by definition. Science becomes bribery by default. Courts become theatre. Elections become rigged performances. The conspiracist does not need to prove each claim; they only need to poison the category of proof itself. Once trust in shared methods collapses, any narrative can reign.

A credible limitation must be stated plainly. Conspiracies do exist. History includes coordinated wrongdoing: Watergate, which began with the break-in at the Democratic National Committee headquarters in Washington, D.C., on 17 June 1972, was real. Investigations confirmed a network of illegal actions connected to the Nixon administration. The danger is not that people suspect hidden coordination; the danger is that they stop distinguishing between bounded, evidence-based conspiracies and totalising, self-sealing ones.

The reconciliation is crucial. Real conspiracies have limits. They involve specific actors, specific actions, and specific evidence trails. They can be investigated and, often, prosecuted. Conspiracy thinking, by contrast, expands indefinitely. It treats lack of evidence as evidence of cover-up. It turns every institution into a mask. It does not merely suspect wrongdoing; it abolishes the possibility of trustworthy knowledge.

This is where the spiritual harm becomes particularly sharp. Conspiracy thinking trains a habit of suspicion that can colonise one's relation to God, neighbour, and self. The believer begins to look for secret messages everywhere, as if divine guidance were primarily a puzzle. They begin to distrust ordinary goods—friendship, counsel, community practice—because "the truth" is assumed to be hidden from the many. The moral life becomes less about love and more about decoding.

The quiet motif threading this discussion is the hidden room. Conspiracy imagines a room behind the room: the meeting behind the meeting, the plan behind the plan. It offers the thrill of access. Yet the hidden room has a cost. If the believer lives mentally in that room, they begin to treat the shared world as a stage set—unreal, manipulated, contemptible.

Once that contempt sets in, community withers. People do not cooperate with those they consider puppets. They do not forgive those they consider agents. They do not listen to those they consider compromised. Even family bonds strain when every disagreement is interpreted as evidence of capture. The behaviour is relational severing. The metric is reduced contact, increased hostility, or constant argument. The implication is loneliness disguised as righteousness.

Modern platforms intensify the pattern. Conspiracy narratives thrive on repetition, novelty, and outrage—conditions that reward engagement. The believer's attention is trained to scan constantly for "clues," making ordinary life feel like a detective story. The mind becomes hypervigilant. Hypervigilance feels like alertness. It is often exhaustion.

And yet the hunger that drives conspiracy is not contemptible. It is a distorted version of something human and, in some sense, honourable: the refusal to accept that suffering is meaningless. The conspiracist cannot bear the thought that tragedies may result from a mixture of error, chance, and systemic fragility. They want intention because intention implies intelligibility. The longing is for a universe that speaks.

The tragedy is that conspiracy answers that longing with a lie that deforms the soul. It replaces meaning with obsession. It replaces trust with suspicion. It replaces humility with omniscience. It offers a counterfeit sacred order: a pattern that is total, thrilling, and morally corrosive.

The forward test is therefore not merely intellectual. When a theory explains everything, when it grants you the comfort of secret sight, when it makes distrust feel like virtue—what will you use to test it? If the hidden room is always available, will you still have the courage to live in the open, where truth is slower, community is messy, and mystery remains part of the human condition?

Discernment in an age of algorithms

In June 2014, researchers at Facebook quietly published a paper describing an experiment conducted on 689,003 users, most of them unaware they were participants. By altering the emotional tone of the News Feed—showing some people more positive posts, others more negative—the team measured small but statistically significant changes in users' own language. The organisation was Facebook. The place was Menlo Park, California. The date matters because the outcome did not concern advertising efficiency or product design. It concerned influence. A pattern-detecting system had nudged the emotional weather of a crowd.

The feed felt neutral.

It was not.

That moment crystallised a shift that had been building for years. Systems designed to recognise patterns in behaviour were no longer confined to laboratories or trading floors. They were mediating attention, belief, and desire at scale. They were not merely reflecting what people wanted; they were shaping what people noticed. The algorithm had begun to act less like a tool and more like an interpreter, selecting which fragments of reality deserved to appear.

The five-word sentence belongs here: *Someone else chooses the patterns.*

The core claim of this section is simple and unsettling. Digital systems now perform a function once reserved for teachers, editors, elders, and clergy: they surface meaning from noise. They decide which signals rise above the background hum of information. They are, in that limited but potent sense, new priests of pattern. The implication is not that they are malicious or mystical, but that they exercise interpretive power. Discernment, once a personal and communal discipline, now unfolds inside architectures designed elsewhere.

To see how this works, watch an ordinary behaviour. A user pauses for two seconds over a video thumbnail. The system records the pause. It compares that micro-behaviour against millions of others. It predicts that a similar clip will hold attention slightly longer. The feed adjusts. This is pattern recognition expressed as code. The name for the process is recommendation optimisation. The explanation is statistical correlation at scale. The implication is cumulative shaping: over time, the feed becomes a personalised corridor, narrowing not by command but by probability.

The abstraction of "attention" becomes visible here as dwell time, scroll depth, and click-through rate. These are not metaphors. They are metrics. They determine which stories, arguments, and images are repeatedly placed before a person's eyes. The system does not ask whether a claim is true, generous, or wise. It asks whether it holds attention. That question is not evil. It is incomplete.

A second named example sharpens the picture. In March 2018, following reporting by *The Guardian* and *The New York Times* on Cambridge Analytica, Facebook acknowledged that data from up to 87 million users had been improperly accessed and used for political profiling. The place again was California, but the effects were global. The outcome was not simply regulatory scrutiny. It was a public recognition that behavioural patterns—likes, shares, network ties—could be marshalled to target beliefs with surgical precision. Pattern detection had crossed into persuasion.

Here the behaviour is micro-targeted messaging. The metric is conversion probability. The implication is asymmetry: one side knows far more about the other than vice versa. Discernment becomes harder when influence is tailored to one's psychological contours.

It is tempting to cast algorithms as villains. That temptation is itself a failure of discernment. Algorithms are not minds. They do not intend. They execute objectives set by humans within economic and cultural constraints. A credible counter-example is worth stating clearly. Recommendation systems have also surfaced neglected art, connected isolated communities, and accelerated scientific collaboration. During the COVID-19 pandemic, preprint servers and social platforms helped researchers share findings rapidly across borders. The same pattern machinery that amplifies misinformation can amplify insight.

The reconciliation lies in recognising function without romance. Algorithms excel at finding correlations in large datasets. They are poor at moral judgement. They optimise for what can be measured, not for what should matter. When treated as neutral mirrors, they quietly govern. When treated as oracles, they mislead. When treated as tools requiring oversight, they can serve.

The quiet motif threading this section is the lens. A lens brings some things into focus while blurring others. No lens is innocent. The question is not whether one uses a lens, but whether one remembers that it is there.

This memory is the beginning of discernment. Discernment, in older traditions, names the practice of distinguishing between signals worth following and those worth resisting. It is not suspicion for its own sake. It is trained attention guided by values. In an algorithmic environment, discernment requires translating ancient questions into modern contexts.

Consider the abstraction of "trust." Trust becomes concrete when one asks which source is repeatedly elevated and which is consistently absent. Trust is enacted when a person chooses to seek primary reporting over viral summaries, or long-form analysis over outrage fragments. The behaviour is deliberate comparison. The metric is diversity of sources encountered. The implication is resilience: a broader informational diet reduces susceptibility to manipulation.

Prediction is the second abstraction made visible. Recommendation engines are prediction machines. They forecast what will keep someone engaged. Human minds are also prediction machines, constantly anticipating what will come next. Problems arise when these two systems lock into feedback loops. The feed predicts what will confirm existing beliefs. The user feels understood. Engagement increases. Alternative perspectives recede. The corridor narrows.

This narrowing is measurable. Studies in network science, including work by Bakshy, Messing, and Adamic at Facebook published in *Science* in 2015, showed that algorithmic ranking and individual choice both contribute to ideological segregation in news exposure. The system and the user co-produce the bubble. Blame is shared. Responsibility remains.

Discernment interrupts this loop not by withdrawal, but by friction. Friction is a choice to slow down. It appears as reading past the headline, checking a claim before sharing, or sitting with discomfort rather than seeking immediate affirmation. These behaviours sound modest. They are not. They resist architectures optimised for speed.

The spiritual cost of ignoring discernment is subtle. When feeds become primary interpreters of reality, moral imagination atrophies. The world appears simpler than it is. Opponents appear flatter. Tragedies become content. The behaviour is reactive consumption. The metric is emotional volatility. The implication is fatigue and cynicism, which are profitable states for systems that monetise attention but corrosive states for communities that require patience.

The lens motif returns when considering authority. In earlier chapters, patterns in stone, sound, and ritual were shown to guide attention deliberately. Cathedrals choreographed awe. Calendars shaped time. These patterns were public, slow, and shared. Algorithmic patterns are private, fast, and opaque. They are customised to the individual rather than binding a community together. The trade-off is efficiency for fragmentation.

No single question can resolve this tension. But discernment begins by noticing which patterns ask for isolation and which invite responsibility. A suggestion that flatters one's fears while discouraging dialogue deserves scrutiny. A recommendation that demands secrecy over accountability deserves resistance. These are not rules. They are habits of mind.

The forward test is therefore lived, not theoretical. When a feed presents a pattern that feels immediately satisfying, that confirms every suspicion, that frames complexity as conspiracy—will you accept the gift without examining the giver? Or will you pause, remember the lens, and ask whether this pattern enlarges your capacity to understand and to care?

Humility before what we cannot map

On 20 March 1968, in Memphis, Tennessee, Martin Luther King Jr stood in the pulpit of Mason Temple and delivered what became known as "I've Been to the Mountaintop." The speech was a response to violence, fatigue, and mounting threats. The outcome was not predictive certainty but a different kind of clarity. King spoke as a man who could not map what was coming and yet refused to pretend otherwise. Ten days later, on 4 April 1968, he was assassinated. The dates do not offer a neat pattern. They refuse it. What remains is the human act of speaking into uncertainty with courage rather than control.

Some events do not submit.

This is not a failure of the universe. It is a limit of our language.

By this point in the book, pattern has been treated as a grammar through which meaning is organised and, at times, disclosed. We have traced how minds stitch, how buildings choreograph awe, how calendars impose rhythm, how algorithms surface correlations, and how false orders seduce. Yet any honest account of pattern must eventually name what pattern cannot do. There are experiences—grief, sudden grace, wonder, moral shock—that resist being translated into geometry without being diminished.

Humility begins where mapping fails.

Mapping is a noble impulse. It is how we make sense, and making sense is how we endure. People who have suffered loss often describe an immediate hunger for explanation: *Why did this happen? What does it mean? What is it for?* These are not childish questions. They are the mind's attempt to restore coherence when reality has been fractured.

Yet grief is not a riddle awaiting solution. It is a wound that changes the body's time.

The behaviour of grief is visible: disrupted sleep, altered appetite, reduced concentration, waves of sorrow that arrive without warning. The metric is not moral improvement or "closure" but persistence over months and years. The implication is that grief does not follow a tidy arc. It loops. It repeats. It returns, often at anniversaries or through sensory triggers—a smell, a song, a familiar light at dusk. People try to map these returns into stages. The stages can be helpful as language, but they can become oppressive when treated as law.

This is one place where modern culture's obsession with explanation becomes cruel. When suffering is framed as a lesson, the bereaved are pressured to translate pain into virtue on demand. When hardship is treated as "for a reason," the unspoken implication is that those who cannot find the reason are failing spiritually.

That implication is false.

Religious traditions, at their most mature, know this. The Book of Job refuses the friends' confident explanations. The Psalms contain lament that does not always resolve into peace. The Christian tradition's focus on the crucifixion insists that the centre of its story is not a neat moral calculus but an event that breaks expectation. The behaviour in these texts is protest and endurance. The metric is fidelity in the absence of understanding. The implication is humility: not everything is legible in the moment, and some things are not legible at all.

If grief resists mapping because it exceeds our capacity, sudden grace resists mapping because it arrives without being earned. It ruptures the causal instinct. A person experiences forgiveness that they did not deserve, help that they did not request, beauty that they did not anticipate. The mind immediately searches for an explanation. *What did I do to deserve this?* But grace, by definition, answers: nothing.

This is why grace feels destabilising. It breaks the pattern of merit.

The visible behaviour is changed action. Someone who receives undeserved kindness may become kinder, not out of obligation but out of altered imagination. The metric is behavioural shift—repaired relationships, softened speech, new willingness to risk trust. The implication is moral expansion: grace can do what punishment cannot. Yet it cannot be engineered. It cannot be summoned by technique. Attempts to map grace into a reliable formula—pray this way, give this amount, follow this ritual precisely—turn it into superstition.

Grace requires surrender, which is deeply inconvenient for controlling minds.

Wonder introduces a third type of resistance. Wonder is not merely pleasure. It is the sensation of encountering something that outruns your categories. A total solar eclipse, for instance, produces a predictable astronomical pattern, yet the human experience often defies prediction: sudden silence, involuntary tears, a feeling of smallness that is not humiliation but relief. The event can be calculated to the second, but the internal response cannot be standardised.

This matters because the modern world often confuses predictability with comprehension. We can forecast storms, model epidemics, and measure neural activity. These are real achievements. But measurement does not exhaust meaning. A brain scan can correlate awe with activity in certain networks. It cannot tell you why awe matters, what it should do to you, or how it should shape your life.

A five-word sentence belongs here: *Numbers cannot carry the weight.*

This is not anti-scientific sentiment. It is intellectual hygiene.

The limit is not that science fails. The limit is that human experience includes layers—moral, relational, existential—that cannot be reduced to pattern without loss. Patterns describe. They do not justify. They can illuminate structure without supplying purpose.

A credible counter-example must be faced. There are times when mapping does help. Epidemiological modelling saves lives. Predictive systems prevent accidents. Psychological frameworks can help people name their experiences and find support. The behaviour is improved decision-making based on data. The metric is reduced harm. The implication is obvious: refusing maps out of romantic attachment to mystery can be irresponsible.

The reconciliation is to distinguish between what can be mapped and what must be honoured as resistant. Humility is not ignorance. It is proportion.

The quiet motif threading this section is the horizon. Maps work within horizons. Beyond the horizon, the map ends. Healthy minds respect this. Unhealthy minds either deny the horizon—claiming total knowledge—or despair at it, treating mystery as meaninglessness. Humility offers a third posture: stand at the horizon without panic, and refuse to pretend you can see past it.

This posture has practical consequences for spiritual life. It protects prayer from becoming a vending machine. It protects theology from becoming conspiracy. It protects community from the cruelty of quick explanations. When someone suffers, humility does not rush to interpret. It stays. It listens. It offers presence rather than a pattern.

Presence is an action. It is measurable. It appears as time spent, meals delivered, messages sent, tears shared, silence kept. The metric is not the elegance of words but the steadiness of companionship. The implication is profound: in the face of what cannot be mapped, the most holy response may be fidelity rather than explanation.

Humility also guards against another temptation: turning the limits of pattern into an excuse for intellectual laziness. Mystery is not a refuge for sloppy thinking. It is a call to careful distinction. Some questions remain open because they are genuinely complex. Others remain open because we have not done the work. Humility does not confuse the two.

A disciplined humility asks: what is knowable here, and by what means? What evidence is available? What remains beyond reach? Which claims are being made with unwarranted certainty, whether by sceptics or believers? It resists both superstition and cynicism because it refuses premature closure in either direction.

This brings the chapter's argument into sharper relief. False orders promise total mapping. They offer to dissolve anxiety with explanation. True order, if it is holy, does not eliminate mystery. It teaches a way to live with it.

The forward test is therefore a matter of character. When life presents you with what resists mapping—a death that makes no sense, a kindness you did not earn, a beauty that stops you—will you rush to force it into a pattern that flatters your need for control? Or will you allow the horizon to remain, and choose a response that is faithful even without a map?

Chapter 9
Living by Pattern

At 6.12 a.m. on a Tuesday in February 2020, the first commuter trains resumed service through Wuhan after weeks of lockdown. The time was published. The routes were specified. The platforms were cleaned and marked. What mattered, though, was not the transport network itself but the re-emergence of routine. People queued. They stood on painted lines. They moved, again, according to a shared rhythm. The restart of daily pattern was treated as a civic milestone, not because trains are sacred objects, but because ordered repetition is how human life becomes livable after disruption.

Lives are shaped long before they are judged.

Most people assume that meaning arrives through large decisions: careers chosen, relationships formed, convictions declared. Yet the deeper architecture of a life is laid down quietly, through habits that repeat without ceremony. Wake times. Meals. Thresholds between work and rest. The way light enters a room in the morning. The path taken to the door. These patterns do not announce themselves as philosophy, but they preach continuously. They instruct the body in what to expect and the mind in what matters.

Pattern, at this scale, is intimate.

The earlier chapters of this book traced pattern across geometry, nature, rhythm, cognition, and belief. They showed how meaning emerges not only from ideas but from repeated forms that train attention and shape behaviour. This chapter turns the lens inward and outward at once. It asks what happens when pattern is not merely observed or interpreted, but inhabited. What does it mean to live as though order matters, not as an abstraction, but as a daily practice?

The modern condition complicates this question. Many people live inside schedules that feel imposed rather than chosen. Notifications fracture time. Work spills into rest. Spaces are crowded with objects that demand attention without offering significance. Community rhythms are fragile, easily disrupted by travel, technology, or exhaustion. Against this background, the idea of *living by pattern* can sound either nostalgic or authoritarian.

It is neither.

Pattern, properly understood, is not rigidity. It is proportion. It is the difference between a life that reacts and a life that responds. A pattern can be strong without being brittle, flexible without being vague. Musical rhythm offers the analogy: without a beat there is only noise; without variation there is only monotony. Meaning lives in the balance.

This chapter is concerned with that balance as it appears in ordinary days.

The first movement considers daily rituals as a form of personal geometry. Just as buildings are shaped by repeated measurements, lives are shaped by repeated actions.

Morning and evening matter because they frame the day. Transitions matter because they signal change. A cup of tea taken at the same time each evening is not trivial; it is a boundary. A short walk before work is not exercise alone; it is orientation. These practices do not require belief to function. They work through repetition, teaching the nervous system what kind of world it inhabits.

Such patterns can be sketched. They can be adjusted. They can be broken and repaired.

The second movement turns to space. Homes and workplaces speak constantly, even when we are not listening. Light directs gaze. Clutter fragments attention. Colour modulates mood. Layout teaches movement. A room can hurry you or slow you down. It can encourage conversation or discourage it. It can suggest care or neglect. None of this requires religious symbols to operate. Geometry preaches quietly, whether acknowledged or not.

To live by pattern is therefore to become literate in environment.

The third movement widens to community. Human beings do not only keep personal time; they share it. Weekly meals, regular gatherings, informal liturgies of greeting and farewell create dependable structures that hold people when individual willpower falters. Communities survive not because everyone feels committed at all times, but because patterns carry commitment through seasons of doubt, fatigue, and change.

Shared rhythm is a form of trust.

The final movement addresses rupture. Patterns break. Illness, injustice, loss, and upheaval can shatter routines that once sustained life. When this happens, the temptation is either to cling rigidly to what no longer fits, or to abandon structure altogether. Neither response heals. Repair requires diagnosis: Which patterns were life-giving? Which were harmful but hidden? What must be rebuilt slowly, with patience rather than force?

Repair is itself a pattern, learned over time.

Throughout this chapter, a quiet motif recurs: the line. A line can divide, but it can also guide. It can be drawn lightly or heavily. It can be erased and redrawn. Lives are not cages of pattern, but pathways traced by repetition. The question is not whether patterns exist, but whether they are chosen with care.

The modern relevance is unavoidable. At a time when many feel overwhelmed by scale—global crises, algorithmic systems, abstract debates—pattern returns agency to the human scale. You cannot control everything. You can shape a morning. You can arrange a room. You can keep a meal. You can rebuild a rhythm after it collapses. These actions do not solve every problem, but they make life inhabitable enough for courage to appear.

The intellectual payoff of this chapter lies in its insistence that meaning is not only something to be understood, but something to be practised. The emotional payoff lies in the recognition that small, repeated acts are not small at all. They are the geometry through which a life takes shape.

The page turns here from interpretation to inhabitation. The patterns explored so far are no longer distant structures in nature, mind, or culture. They are waiting at the scale of days. The question that remains is simple, and demanding: if pattern truly carries meaning, how will you choose to live inside it?

Daily rituals as personal geometry

On 12 September 2019, at 5.45 a.m., Haruki Murakami ended his morning run along the Charles River in Cambridge, Massachusetts. He showered, ate a simple breakfast, and sat down to write for exactly four hours. He has described this routine publicly, repeatedly, with almost pedantic precision. The outcome is measurable: more than a dozen novels produced across decades with striking consistency. The interest here is not literary celebrity, but geometry. Murakami's life, at least in its working hours, is drawn with straight lines and repeated angles. The day is not left to chance. It is measured.

A life acquires shape before it acquires meaning.

Most people imagine routines as neutral scaffolding, necessary but uninteresting. In practice, routines are formative forces. They decide what receives attention when the mind is most pliable, what the body expects at moments of transition, and how energy is conserved or wasted across a day. The geometry of a life is not found in its aspirations but in its repetitions.

Consider the morning. The first hour after waking is neurologically distinctive. Cortisol levels rise sharply; attention is easily captured; habits dominate over deliberation. What happens here is rarely chosen consciously, yet it sets the tone for everything that follows. Checking a phone within minutes of waking produces a measurable effect: studies using experience sampling methods, such as those summarised by Rosen, Lim, Smith, and Smith in 2014, show increased distraction and anxiety across the subsequent morning. The behaviour is visible. The implication is structural. The day begins fragmented.

By contrast, a repeated action that anchors the morning—tea prepared in the same cup, a short walk, a fixed period of silence—functions as a stabilising line. This is not mysticism. It is training. The nervous system learns what to expect. The body settles into proportion. The mind is given a boundary.

Geometry teaches through constraint.

Routines operate most powerfully at thresholds. The movement from sleep to waking, from work to rest, from public to private space. In architecture, thresholds are rarely accidental. Doors, porches, courtyards, and vestibules exist to prepare the body for change. Daily rituals serve the same function internally. A brief act at the end of work—a notebook closed, a lamp switched on, shoes removed—

marks a transition more effectively than intention alone. Without such markers, roles bleed into one another. Work colonises rest. Rest loses its depth.

This is visible in behaviour. Time-use surveys from the UK Office for National Statistics show that people who report no clear boundary between work and leisure spend longer working hours overall and report lower life satisfaction, even when income is controlled. The pattern, not the content, does the damage.

Evening rituals matter for similar reasons. Sleep quality correlates strongly with regular pre-sleep behaviour. The National Sleep Foundation's 2017 Sleep in America poll found that individuals with consistent evening routines fell asleep faster and reported fewer awakenings. Again, the mechanism is not belief but repetition. The body responds to pattern.

Routine, then, is not the enemy of freedom. It is its precondition.

A common objection arises here. Routines can harden into ruts. They can become lifeless, oppressive, or exclusionary. This is a real risk. Monastic rules, factory shifts, and domestic expectations have all been used to constrain rather than liberate. The counter-example matters. Geometry can imprison as well as guide.

The reconciliation lies in proportion. Healthy routines are characterised by three features: they are simple, they are repeatable, and they are revisable. Simplicity prevents overload. Repeatability ensures reliability. Revisability allows growth. When any of these fail, routine degrades into rigidity.

Take the practice of daily prayer across traditions. In the Benedictine Rule, written in the sixth century in Italy, prayer is fixed to hours, but the content varies. The geometry is stable; the experience is not identical. This balance has sustained communities for centuries. Where prayer becomes mechanically fixed without room for variation, participation often collapses. The pattern remains, but meaning drains away.

The same principle applies in secular lives. A daily walk that becomes impossible to alter breeds resentment. One that can shift in length or route retains life.

The invitation implicit in personal geometry is not imitation but design.

To sketch one's own daily patterns is to observe without judgement. What repeats? What drains energy? What restores it? These questions are diagnostic, not moral. A pattern that looks virtuous may be unsustainable. A pattern that looks trivial may be essential. The metric is consequence. Does this repetition widen attention or narrow it? Does it steady the body or agitate it?

Trust, often treated as an abstraction, appears here as a behaviour: returning to the same stabilising practice even when motivation wavers. Efficiency appears as a metric: reduced decision-making at points of fatigue. Risk becomes visible as the choice to leave certain hours unstructured, allowing for surprise without chaos.

The quiet motif running through this reflection is the line. A line drawn too rigidly cracks under pressure. A line drawn too vaguely guides nothing. The art lies in drawing lines that can be redrawn.

Modern life exerts pressure against such drawing. Flexibility is prized, but often confused with formlessness. Digital tools promise customisation while eroding rhythm. Notifications arrive without regard for threshold. Days become collections of fragments. In this context, choosing routine is a countercultural act.

It is also a spiritual one, whether named as such or not.

Spiritual traditions have long understood that character is trained indirectly. You become patient by keeping time with others. You become attentive by repeating acts that require attention. You become generous by structuring life so that generosity is possible. None of this depends on constant feeling. It depends on geometry.

Yet humility remains necessary. Not every life permits the same patterns. Illness, caregiving, poverty, and crisis disrupt routine through no fault of the individual. To moralise routine is to misunderstand it. Geometry describes structure, not virtue.

The closing question, then, is not whether routines are good or bad, but whether the lines you draw each day are drawn deliberately. If the geometry of a life is already forming through repetition, the unresolved tension is simple: will you choose the shape, or continue to inhabit one drawn by accident?

Designing spaces that preach quietly

In October 1984, in the Hawthorne Works outside Chicago, the sociologist and organisational researchers who were studying factory productivity published a final report that turned a simple insight into a lasting term: workers change their behaviour when they know they are being observed. The "Hawthorne effect" is often treated as a caution about experiments. It is also a clue about rooms. People adapt to environments, not only because of rules, but because of cues. Light, sightlines, noise, and layout tell a body how to behave before a supervisor speaks.

Space instructs in silence.

Most of us treat homes and workplaces as neutral containers for life. We move furniture when it becomes inconvenient, buy a lamp when a corner feels dark, clear a desk when deadlines loom. Yet environments are not passive. They teach the nervous system what to expect. They reward certain behaviours and punish others. Over time, they form habits as surely as any schedule.

The central claim is disarmingly practical: design is theology by other means. Even in secular rooms, built form expresses beliefs about attention, community, and what counts as valuable work. A room may preach speed.

Another may preach patience. A corridor may preach hierarchy. A kitchen table may preach mutuality. None of this requires slogans. Geometry does the speaking.

Begin with light. Light is not merely illumination; it is a regulator. The body's circadian system responds strongly to brightness, timing, and colour temperature. When daytime spaces are dim and evening spaces are saturated with bright blue light from screens, sleep quality degrades, mood shifts, and attention becomes brittle. These effects are measurable. The UK's National Health Service and the Scientific Advisory Committee on Nutrition have both highlighted the importance of regular sleep and light exposure for health, and the broader scientific literature on circadian entrainment is extensive. The behavioural consequence is visible in late-night scrolling and early-morning fatigue. The implication is moral as well as physiological: tired people make poorer decisions, show less patience, and struggle to sustain reflective attention.

A room that "preaches quietly" often begins by telling the body when it is day and when it is night. That is a form of care, not décor.

Clutter is the second preacher. Clutter is not simply mess; it is visual noise. Researchers in environmental psychology have repeatedly found that complex, highly cluttered environments increase cognitive load—the brain must filter more stimuli, leaving less capacity for sustained focus. The behaviour is fractured attention: more task-switching, more avoidance, more low-grade agitation. The metric is time lost to reorienting—finding objects, clearing surfaces, shifting piles. The implication is predictable: the room becomes a machine for distraction.

Yet here too a limitation matters. Some people thrive amid "organised chaos." Creative studios, workshops, and family homes can be busy and still nourishing. The counter-example is real. Clutter only becomes harmful when it produces continual micro-decisions: where to put things, what to ignore, what demands action. The reconciliation is not minimalism as an ideology, but legibility. A space should make its priorities easy to read.

Legibility is the spiritual virtue of architecture.

Colour shapes mood in quieter ways. This is not the simplistic claim that blue "calms" and red "energises." The effect of colour depends on context, saturation, and personal history. Still, design research and studies of workplace environments suggest that high-contrast, saturated visual fields can increase arousal, while softer palettes can reduce it. The behaviour is subtle: people speak louder in bright, echoing rooms; they linger longer in warmer, absorbent ones. The metric is dwell time and conversational tone. The implication is communal: the room either supports presence or accelerates departure.

Layout is where geometry becomes explicit. In workplaces, open-plan offices became fashionable as symbols of collaboration. The belief was moralised: walls are silos; openness is trust. Yet research has complicated this story. A widely cited study by Ethan Bernstein and Stephen Turban published in *Philosophical Transactions of the Royal Society B* in 2018 reported that moving to open-plan offices reduced face-to-face interaction and increased electronic communication. The behaviour was measurable: fewer in-person conversations, more messaging. The implication was paradoxical: spaces designed to encourage community sometimes produce withdrawal because they reduce privacy and increase self-consciousness.

The room preached exposure. People responded with retreat.

Homes have analogous dynamics. A kitchen that flows into a living area invites conversation and shared labour, but it can also abolish retreat, leaving no place for quiet. A bedroom that doubles as office collapses the boundary between work and rest. The behaviour is persistent mental activation. The metric is difficulty switching off. The implication is erosion of recovery.

This is why "sacred geometry in ordinary rooms" should not be understood as exotic symbols or esoteric ratios. It begins with threshold and orientation. Where does the eye go when one enters? Where does the body naturally pause? What is centred and what is peripheral? These are design questions, but they are also questions about attention and values.

A five-word sentence belongs here: *Centres make lives feel held.*

Centres need not be grand. A dining table that is cleared each evening becomes a communal centre. A chair by a window becomes a contemplative centre. A shelf with a few chosen objects becomes a centre of memory. The behaviour is regular return. The metric is frequency of use. The implication is stability: a centred room reduces the frantic search for stimulation because it offers a place to settle.

Transitions matter, too. Earlier we noted that daily rituals mark thresholds. Rooms can do this physically. A small mat by the door, a hook for keys, a lamp switched on at dusk—these are not trivial. They reduce friction. They prevent the mind from bleeding one mode into another. The abstraction of "rest" becomes a visible practice: shoes off, coat hung, light softened. The metric is fewer unresolved tasks carried into evening. The implication is deeper rest.

Sound is often ignored in discussions of space, yet it may be the most morally consequential element. Hard surfaces amplify noise, increasing stress and reducing capacity for patient conversation. Soft furnishings absorb noise, allowing voices to become gentler. The behaviour is measurable: interruptions increase when background noise is high; listening declines. The implication is relational. Many arguments begin not with disagreement but with exhausted auditory processing.

A room can teach people to shout.

Workspaces "preach" through hierarchy embedded in layout: who gets the window, who sits near power, who must pass whom to reach a bathroom. These are not merely logistical choices; they form social patterns. The behaviour is deference or dominance. The metric is who speaks most in meetings, who interrupts, who lingers in shared areas. The implication is cultural: space stabilises status.

In homes, the same principle appears in who controls common areas, whose belongings are allowed to spread, whose needs determine quiet. A "shared" room is often shared in theory and contested in practice. Designing spaces that preach quietly therefore includes moral negotiation. It is the recognition that geometry is communal.

Here the motif of preaching returns with sharper teeth. Preaching is not always kind. A room can preach that you are behind, that you never finish, that you must always be available. Many modern work tools and environments have been designed to maximise output, not dignity. The result is not only burnout but a disordered inner life. People become unable to rest without guilt because the room never signals permission.

Discernment, applied to rooms, asks: what does this space train me to expect of myself and of others?

This can be tested without grand theory. Measure how often you move in a room without purpose. Notice whether you can read without checking a device. Observe whether conversations in a space tend to deepen or stay superficial. These are behavioural indicators. They reveal what the room is teaching.

The religious analogy is not decorative; it is precise. Traditional sacred spaces manage attention through light, axis, and repetition. The nave draws the body forward. The mihrab orients prayer. The mandala centres contemplation. Ordinary rooms can borrow the principle without borrowing the symbol. A workspace can be oriented towards one task at a time rather than perpetual interruption. A living room can be arranged to face people rather than screens. A corridor can be cleared so that coming home feels like entering, not stumbling.

None of this requires expense. It requires intention.

The unresolved tension is that modern life often forces people into spaces they did not choose: rented rooms, shared houses, offices designed by distant committees. Perfect control is rare. Yet even within constraints, small choices alter geometry. A lamp can soften a corner. A chair can claim a refuge. A cleared surface can become an altar to attention, whether or not one uses religious language.

The forward test is therefore simple and difficult. When you step into your home or workplace, what sermon does the space deliver before you speak a word—and will you accept that sermon, or redraw the room's geometry until it teaches something truer?

Community patterns, liturgy, and shared time

On 11 March 2020, in Rome, Pope Francis stood alone in a rain-darkened St Peter's Square to deliver an *Urbi et Orbi* blessing during the first wave of COVID-19. The place was almost theatrically empty, the colonnades framing absence as much as presence. The outcome, however, was not emptiness. Millions watched across continents. A liturgy was performed without a congregation in the square, yet with a congregation dispersed through screens and silence. The scene exposed something that modern life often disguises: communities do not only gather in space. They gather in time.

Time is the true meeting place.

When people speak of belonging, they usually describe relationships, beliefs, or shared history. Yet belonging is also a schedule. It is the repeated return to the same hour, the same meal, the same day of the week. Without these returns, relationships become occasional, beliefs become private opinions, and shared history becomes nostalgia. Community is not sustained by intensity. It is sustained by rhythm.

Liturgy is simply the name given to this rhythm when it is deliberate.

Many modern readers hear "liturgy" and imagine formal worship: robes, scripted prayers, incense, hymns. That is one species. But liturgy is broader than that. It is any repeated, communal pattern that structures attention, action, and meaning. A weekly family dinner is a liturgy. A regular meeting that begins with the same question—*how are you really?*—is a liturgy. A neighbourhood that gathers each Saturday morning to clean a shared space is a liturgy.

The forms are informal, but the function is identical: to hold people together through time.

This holding is not sentimental. It is structural.

Consider what weekly rhythms accomplish. They reduce the burden of constant negotiation. A group does not have to reinvent its togetherness each week; it returns to it. That return is a behaviour. The metric is attendance over months, not emotional tone on any given day. The implication is resilience: when individual motivation dips, the pattern carries the relationship.

Sociological research has long observed that regular communal participation correlates with social support and wellbeing, though the mechanisms vary. Robert Putnam's *Bowling Alone* (2000) remains a widely cited account of declining civic participation in the United States, linking reduced involvement in clubs, churches, and community organisations to weaker social capital. The thesis has been debated, but the basic observation stands: when shared rituals decline, loneliness rises, and trust becomes harder to sustain.

Trust is not an inner feeling. It is a repeated practice.

This is where shared meals matter. Eating together is one of the oldest human technologies for belonging. It coordinates time, attention, and reciprocity in one act. Someone prepares food. Others arrive. Conversation unfolds within a bounded duration. The meal ends. People leave, slightly altered. The behaviour is visible: faces oriented towards one another, hands occupied, pauses naturally punctuating speech. The metric is frequency and duration. The implication is intimacy without interrogation: people can be together without demanding constant verbal performance.

Religious traditions formalise this instinct because it works. The Christian Eucharist, the Jewish Sabbath meal, the Muslim iftar during Ramadan, the Sikh langar offered in gurdwaras—all treat shared eating as both sustenance and sign. Yet even stripped of theological language, the structure remains potent. A meal is a recurring centre.

A five-word sentence belongs here: *We become what we repeat.*

Repetition is often caricatured as dullness. In communal life, repetition is mercy. It means you do not have to earn your place anew each week. You can return tired. You can return grieving. You can return distracted. The pattern receives you.

This receiving is one reason liturgy holds people through change. Life changes faster than relationships can sometimes adapt. Jobs shift. Children grow. Illness arrives. Grief interrupts. Without a shared rhythm, each change demands renegotiation: *When can we meet? Do we still belong? Are we still close?* A stable communal pattern reduces the number of decisions required to remain connected. The calendar becomes a covenant.

Yet there is a limitation, and it must be faced. Liturgies can ossify. They can become performative, excluding those who do not know the cues. They can enforce conformity under the banner of tradition. Anyone who has sat through a meeting where ritualised politeness masks hostility knows that repetition can be used to avoid truth rather than to sustain it.

The counter-example is real: a community can keep perfect rhythm while losing love.

The reconciliation lies in distinguishing form from purpose. A living liturgy is not merely repeated behaviour; it is repeated behaviour ordered toward a shared good. The behaviour includes welcome, listening, and the distribution of responsibility. The metric is not only attendance but participation: who speaks, who hosts, who is noticed, who is carried when they cannot contribute. The implication is justice within community. Without this, rhythm becomes a metronome that measures decay.

Informal liturgies reveal this quickly. Consider workplace rituals. Some organisations begin each week with a "stand-up" meeting. In theory, it produces alignment. In practice, it can become a daily performance of busyness. The behaviour is status reporting. The metric is speed and visibility. The implication is anxiety: people optimise for appearing productive rather than doing meaningful work. That is a liturgy, too—just not a holy one.

A healthier communal rhythm looks different. It makes room for truth. It includes some form of check-in that is not purely instrumental. It creates predictable opportunities for gratitude and repair. It marks endings as well as beginnings. These are not decorative niceties. They are structural supports for trust.

Shared time also trains moral imagination. When people meet regularly, they are exposed to one another's ordinary reality. They see each other age, fail, recover, and change their minds. This gradual exposure makes caricature harder. It counteracts the flattening effects of distant opinion and digital abstraction. You cannot easily demonise someone whose life you have witnessed in seasons.

The behaviour is sustained proximity. The metric is years, not posts. The implication is compassion, because complexity becomes unavoidable.

This point has particular force in a culture saturated with personalised media. Digital platforms allow us to curate communities of agreement, minimising friction and maximising affirmation. Shared time in embodied communities works differently. It forces people to endure disagreement without immediate exit. It trains patience. It requires forgiveness. These are not optional virtues if a community is to persist.

Of course, not every community is safe. Some patterns should be broken. Earlier chapters examined false orders and the dangers of rigid, self-sealing systems. The same is true here. A "community rhythm" that consistently silences some members or normalises harm is not a stabilising pattern; it is a cage. Discernment remains necessary.

Yet the need for discernment does not cancel the need for rhythm. It heightens it.

A quiet motif threads through this discussion: the table. The table is a physical object, but also a symbol of shared time. It is where the weekly meal happens, where the meeting begins, where hands pass bread or papers. A table can include. It can exclude. It can be rearranged. The table is not sacred because of wood or polish. It is sacred because it gathers attention.

Communal patterns are ultimately patterns of attention. What does a community notice together? What does it ignore together? What does it celebrate? What does it mourn? Liturgies—formal or informal—answer these questions by repetition. They teach a group what is real.

The forward test is therefore not whether you belong to a community in name, but whether you share time in a way that can withstand change. If your life is scheduled to the minute yet lacks dependable return to others, what exactly is holding you? And if you do return—weekly, monthly, at some chosen rhythm—does the pattern enlarge trust and care, or does it merely keep the machinery running?

Repairing broken patterns in life

In September 2005, in the Lower Ninth Ward of New Orleans, residents returned after Hurricane Katrina to streets where addresses had become guesses and landmarks had vanished. The organisation coordinating much of the early rebuilding was not a single church or agency but a swarm of groups, among them Habitat for Humanity, which expanded its work in the city in the months that followed. The concrete outcome was visible in timber frames rising where waterlines still stained walls. Yet the deeper rebuilding was less photogenic. It was the slow reassembly of pattern: school runs, Sunday gatherings, neighbourly exchanges, the ordinary repetition that makes a place feel inhabited rather than merely survived.

Disaster breaks geometry first.

The collapse of pattern is often discussed in psychological terms—trauma, anxiety, burnout—and in material terms—lost income, damaged homes, disrupted services. These descriptions matter, but they can miss the structural truth: crisis is a sudden rupture in the repeated forms that carry life. When the familiar sequences vanish, people do not merely lose comfort; they lose orientation. The body does not know when to rest. The mind does not know what to expect. Communities lose the shared rhythm through which trust is enacted.

The same is true of injustice, though it operates less like a storm and more like a slow erosion. Injustice is a pattern that harms by repetition. It becomes normal precisely because it is normal. It teaches people to anticipate disrespect, exclusion, or danger as part of the daily order of things. When injustice is challenged, the resulting conflict can feel like collapse, but it is often the first honest diagnosis.

Repair begins with naming what has broken and what was broken already.

The temptation in crisis is to reach for speed. People want the old patterns back quickly, because pattern is safety. Yet speed can be a kind of denial. Some patterns should not return. An exhausted work rhythm that led to collapse should not be reinstated unchanged. A communal culture that protected abusers should not be "restored." A family pattern of silence around conflict should not be revived in the name of peace.

This is why repair requires discernment, not nostalgia.

Consider the behaviour of a person after a major disruption—bereavement, redundancy, illness. At first, routines can feel impossible. Small tasks acquire unexpected weight. Sleep fractures. Concentration blurs. The metric is not willpower but baseline functioning: whether one can eat, wash, move, and communicate reliably. The implication is often misread. People assume they are failing because they cannot "get back to normal." In reality, normal has been replaced. The nervous system is recalibrating. Demanding immediate coherence can prolong suffering.

The most humane repair begins with minimal, repeatable structure. This is not self-improvement culture. It is stabilisation.

Here a limitation must be stated plainly. There are contexts where minimal structure is not enough, where professional support and systemic intervention are necessary. Abuse, severe mental illness, chronic poverty, or institutional discrimination cannot be solved by personal routine alone. To suggest otherwise would be cruel. The reconciliation is to treat personal pattern as one layer of repair among others, not a substitute for justice or care.

Repair is layered work.

In communities, diagnosis follows the same logic. When patterns collapse—after a scandal, a financial shock, a violent incident—the group often argues about causes. Some want to blame individuals; others blame systems; still others spiritualise the crisis as punishment or test. These interpretations can become secondary damage. The more useful approach is to map behaviour without moral panic. What did people do repeatedly that produced vulnerability? What warning signs were ignored? What incentives shaped decisions? What voices were missing at the table?

These questions translate abstraction into evidence.

A community's "trust" becomes visible as reporting behaviour: do people raise concerns early, or only when collapse forces confession? "Efficiency" becomes visible as meeting structure: are decisions made transparently, or in backchannels? "Risk" becomes visible as redundancy: does the group rely on one charismatic leader, one donor, one organiser, or does it distribute responsibility?

Distribution matters. Systems that depend on a single node fail catastrophically when that node breaks.

A named example beyond disaster clarifies this. In June 2017, after the Grenfell Tower fire in London, survivors and local residents faced not only grief but disorientation: housing uncertainty, bureaucratic confusion, mistrust of authorities. Community organisations and resident groups stepped into the gap, providing mutual aid, advocacy, and practical support. The public inquiry and subsequent debates exposed how long-standing patterns—regulatory decisions, maintenance practices, and the treatment of residents' warnings—had created vulnerability. The fire was an event. The conditions were a pattern.

Repair, in such cases, cannot be a return. It must be a reconfiguration.

This is the heart of the matter: broken patterns often reveal hidden patterns. Crisis is diagnostic. It shows what was fragile, what was unjust, what was dependent on luck rather than structure.

The quiet motif threading this section is the seam. Repair is not the erasure of a tear; it is the making of a seam. A seam is visible. It holds because it acknowledges the break. It is stitched, not wished.

Seams are made through patient repetition. This is why rebuilding healthier structures often feels unimpressive. It involves meetings that are shorter and more consistent. It involves clearer roles. It involves agreed thresholds for rest.

It involves mechanisms for feedback. It involves rituals of apology and restitution that do not depend on mood.

In religious communities, such seams are often ritualised. Confession, reconciliation, lament, and communal prayer function as repair patterns. They provide language and timing for what would otherwise be chaotic: naming harm, seeking forgiveness, re-entering relationship. Even when a community is not explicitly religious, analogous practices exist. A workplace that institutionalises post-mortems after failures is practising a secular liturgy of repair. The behaviour is structured reflection. The metric is recurrence: whether learning happens after each incident rather than only after catastrophe. The implication is resilience.

Yet repair demands something modern culture finds difficult: slowness. After crisis, people want a statement, a policy, a redesign. These may be necessary. But without repeated enactment, they remain theatre. A healthier structure is one that can be lived. It must survive fatigue, conflict, and ordinary forgetfulness.

A five-word sentence belongs here: *Repair is repetition with care.*

Care appears in small decisions. Who gets listened to first? Who bears the burden of rebuilding? Who is protected? Who is asked to perform forgiveness prematurely? A community can rebuild physically while leaving relational damage untreated. It can declare itself healed while people remain unsafe. The metric that matters is participation without fear.

This is where injustice complicates repair. When patterns collapse because a marginalised group refuses further harm, the "old normal" is often defended as stability. The community must decide whether it values peace or truth. Repair that restores an unjust pattern is not repair. It is reinforcement.

The personal analogue is equally sharp. Many individuals rebuild their lives after crisis by returning to the patterns that contributed to collapse: overwork, avoidance, compulsive distraction, isolation. These patterns are familiar, and familiarity impersonates safety. Repair requires a different kind of courage: the willingness to build unfamiliar rhythms slowly.

Slow building is not indecision. It is design under constraint.

The most difficult part of repair is often the transition period, when the old pattern has been broken but the new one is not yet established. This liminal space can feel like failure. It is not. It is the seam being stitched. Communities in this phase are vulnerable to false orders—the quick fix, the scapegoat, the charismatic saviour. Individuals are vulnerable to the same: impulsive commitments, rigid vows, dramatic reinventions that collapse within weeks.

Discernment asks for tests rather than declarations. Does this new pattern reduce harm? Does it distribute responsibility? Does it create time for rest? Does it allow truth to be spoken without retaliation? These are behavioural criteria, not slogans.

The unresolved tension at the end of repair is always the same. After crisis, a person or community can choose comfort or transformation. Comfort seeks to feel normal again as quickly as possible. Transformation accepts the seam, learns from the break, and commits to a structure that can hold in the next storm. The forward test is therefore simple, and it will not stay theoretical: when the first signs of stability return, will you rush to re-inhabit the old geometry, or will you keep stitching until the new pattern can truly bear weight?

Chapter 10
The Future of Holy Pattern

In March 2021, a small chapel appeared inside a video game. Built within *Minecraft* by a group of architecture students in Copenhagen, it was not a novelty map but a functioning space for quiet gathering during pandemic lockdowns. Participants logged in from different countries, moved their blocky avatars into a simple nave, and sat in silence together. There was no sermon. No spectacle. The outcome, reported by the organisers in interviews that year, was unexpected: people stayed longer than planned. They returned. Some described the space as calmer than their physical surroundings. A sacred form, translated into pixels, had held attention.

The scene raises a question that would have sounded absurd a generation ago: when space becomes code, what happens to holiness?

This final chapter stands at a threshold. Everything so far has argued that pattern is not decoration but orientation; not an optional flourish but a deep grammar through which meaning becomes livable. We have traced pattern from stone temples to daily habits, from neural prediction to communal repair. Now the question sharpens. If pattern has always been the medium through which human beings encounter order, responsibility, and transcendence, what happens as the mediums themselves change?

The future of holy pattern is not about novelty. It is about fidelity under new conditions.

Digital worlds are the most obvious testing ground. Virtual and augmented environments already structure attention with a precision no cathedral architect could have imagined. Interfaces decide where the eye goes, what the hand does, how time feels. Risk follows immediately. When pattern is optimised solely for engagement metrics, it trains restlessness. When symmetry serves persuasion rather than contemplation, it becomes manipulation. Yet possibility follows just as closely. The same tools can create spaces that slow perception, invite reflection, and teach restraint. The question is not whether sacred design can survive digitisation, but whether designers and users will recognise that pattern carries ethical weight even when it is weightless.

This matters now because more of life is mediated. Meetings, friendships, rituals of work and rest increasingly occur inside designed systems whose values are implicit rather than declared. A feed is a liturgy. A notification cadence is a theology of urgency. To pretend neutrality here is to abdicate responsibility. This chapter asks what it would mean to bring the long wisdom of sacred geometry—thresholds, proportions, rhythms—into spaces built of light and logic.

The inquiry widens further when science enters the frame. Contemporary physics and cosmology are not returning to pre-modern mysticism, but they are circling old questions with new tools. Why is the universe describable by mathematics at all? Why do abstract equations written on paper predict the behaviour of stars and particles with such accuracy? Physicists from Eugene Wigner onwards have called this intelligibility "unreasonable." Theologians, philosophers, and scientists now meet in shared perplexity rather than rivalry. Consciousness, emergence, order: these are not sectarian concerns. They are boundary questions, where explanation thins and wonder presses in.

This chapter does not collapse science into spirituality or dress equations in incense. It holds a more demanding posture: that humility before intelligibility itself may be the most honest common ground. Pattern here is not proof of purpose, but invitation to responsibility. If the universe can be known, even partially, then how knowledge is used becomes a moral question.

Which brings us to ethics, the most practical future of all. Moral life has always relied on pattern, even when it pretends not to. Laws visualise balance. Courts arrange opposing claims across space. Restorative justice circles reconfigure relationships rather than merely punishing breaches. Systems thinking maps feedback loops to reveal where harm concentrates. These are not metaphors; they are diagrams of responsibility. To speak of ethics written in pattern is to insist that justice is not only a rule but a shape, and that changing outcomes requires changing arrangements.

This has urgent relevance in a world of complex systems—economic, ecological, technological—where intention alone no longer guarantees good results. Pattern literacy becomes a civic virtue. The ability to see how actions repeat, amplify, or cancel each other is no longer academic. It is survival work.

Yet the chapter does not end in abstraction. It gathers its arguments into a final, personal claim. If patterns organise worlds, then lives are motifs. Each person and community repeats certain forms—ways of speaking, moving, deciding—that echo outward. Some motifs harmonise. Others distort. None are neutral. To live consciously, acknowledging oneself as a recurring figure within a larger design, is not grandiosity. It is restraint. It recognises scale without denying agency.

The future of holy pattern is therefore not a prediction. It is a posture. It asks whether, amid accelerating systems and proliferating designs, we will choose patterns that train attention, distribute care, and leave room for what cannot be mapped. The pages ahead explore where this choice is already being made, wisely or otherwise, and what it might mean to make it deliberately.

The book has traced lines and cycles, thresholds and returns. This final chapter asks a simpler, harder question: when the patterns of tomorrow are being drawn now, often invisibly, who will notice—and who will decide which forms are worthy of repetition?

Digital worlds and sacred design

In April 2020, as churches across Europe closed their doors, a small Lutheran parish in Aarhus, Denmark, authorised something unprecedented. With the blessing of its clergy, a volunteer collective rebuilt the church interior inside *Minecraft*, opening it to parishioners during lockdown. The date matters. April 2020 marked the first Easter in living memory when physical gathering was impossible for millions. The outcome was concrete: attendance exceeded the usual Sunday count, and participants reported staying in silence together for extended periods rather than dispersing after prayers. A sacred form, rendered in cubes and code, had altered behaviour.

This was not nostalgia. It was design.

The experiment exposes a problem that can no longer be avoided. Virtual and augmented environments are no longer marginal spaces of entertainment. They are becoming primary theatres of attention, work, and social presence. If sacred architecture has historically shaped posture, gaze, and breath through stone and proportion, what happens when those shaping forces are written instead in pixels, shaders, and interaction rules? The question is not whether digital worlds will host spiritual life, but whether they will do so carelessly or with discernment.

The first claim is visible before it is named. Enter any large virtual platform and notice the body. The shoulders rise.

The eyes flick. Time compresses. This is not accidental. These spaces are optimised for engagement measured in clicks, dwell time, and behavioural prediction. Sacred design, by contrast, has always optimised for *thresholds*. The narthex slows entry. The courtyard mediates noise. The aisle disciplines pace. Naming the difference clarifies the issue: digital environments currently privilege capture over contemplation. The implication is stark. When sacred forms are imported into such systems without altering their underlying metrics, they risk becoming decorative skins over restless engines.

Yet the same tools can be used otherwise. Consider the virtual reconstruction of Notre-Dame de Paris commissioned after the 2019 fire, drawing on laser scans produced by art historian Andrew Tallon before his death in 2018. The digital model, used by restoration teams and made available for immersive viewing, does something subtle. It restores scale. Users report instinctively slowing movement when navigating the nave. The reason is not sentiment but geometry. Proportions remain legible even when gravity disappears. Pattern still teaches the body how to behave.

This shows the second claim. Pattern survives translation when structure, not surface, is preserved. Sacred geometry is not an aesthetic overlay; it is an algorithm for attention. When virtual environments honour ratios, axiality, and rhythm rather than merely visual motifs, they reproduce effects familiar from physical temples. Gaze lifts. Pace steadies. Silence becomes possible. The implication is practical. Designers of digital spaces cannot outsource meaning to symbols alone. They must decide how users move, pause, and return.

The risks sharpen when augmentation enters daily life. Augmented reality overlays information directly onto streets, rooms, and faces. In May 2023, Apple announced its Vision Pro headset in Cupertino, presenting spatial computing as the next interface paradigm. The promise is seamless integration of digital pattern into physical space. The hazard is saturation. When every surface can speak, nothing listens. Sacred design has always rationed information. Iconography appears where attention has been prepared. Thresholds protect the centre. Without such restraint, augmented worlds risk flattening hierarchy, dissolving the distinction between signal and noise.

A counter-example deserves attention. The Benedictine abbey of Münsterschwarzach in Germany experimented quietly during the pandemic with a simple augmented application: visitors could, at specific points, see brief illuminated lines of chant notation hovering near the choir stalls, triggered only during prescribed hours. No sound played. No interaction followed. The effect, reported in 2021 interviews, was not distraction but orientation. The augmentation did not add content so much as reveal rhythm already present. This suggests reconciliation. Digital layers can serve sacred ends when they are governed by liturgical time rather than constant availability.

The motif threading through these cases is the *threshold*. Sacred spaces, physical or digital, work when entry is marked, not frictionless. Virtual reality prides itself on immersion without resistance. Sacred design insists on transition. The tension between these logics defines the future.

A further claim emerges when we examine community. Digital sacred spaces often promise accessibility. Anyone can enter. No travel. No exclusion. The virtue is real. Yet communities are not formed by access alone but by shared discipline. Medieval pilgrimage routes limited speed. Monastic schedules constrained choice. In digital worlds, where movement is instantaneous and exit costs nothing, pattern must work harder. Without shared rhythms—times, gestures, repetitions—spaces dissolve into private consumption.

This is visible in the failure of many online worship experiments between 2020 and 2022. Streams drew viewers but not participants. Metrics spiked, then declined. By contrast, smaller communities that imposed pattern—fixed start times, collective silences, repeated forms—retained engagement. The abstraction is *belonging*. The metric is return rate. The implication is clear. Sacred design in digital environments must privilege recurrence over novelty.

A limitation must be faced honestly. Not all elements of sacred space translate. Smell, weight, fatigue, and unchosen proximity resist simulation. To deny this is to romanticise technology. Digital sacred design cannot replace physical gathering. It can only supplement it, or serve those excluded by circumstance. The reconciliation lies in humility. Virtual spaces should be designed as *chapels,* not cathedrals: focused, limited, aware of what they cannot carry.

Ethical stakes follow inevitably. Algorithms already shape belief by surfacing patterns of preference. Recommendation engines decide which voices repeat. In this sense, platforms already function as priests of pattern, mediating visibility and invisibility. Sacred design demands accountability here. Who decides which paths are emphasised? Which returns are encouraged? A digital labyrinth that always leads to outrage is not neutral architecture. It is catechesis by code.

The future question, then, is diagnostic rather than predictive. When a digital sacred space calms attention, does it do so by slowing the body or by numbing it? When it gathers people, does it ask anything of them beyond presence? When it repeats forms, do those repetitions train patience or dependence? These are not technical queries. They are moral tests embedded in design decisions.

The chapter ends without resolution because the work is unfinished. Pixels will continue to host pattern. Code will increasingly shape devotion, reflection, and moral imagination. The choice is whether these patterns will be shallow, optimised for extraction, or deep enough to bear silence, limit, and return. Sacred design has always been an argument about what deserves repetition. The digital future will amplify that argument, not replace it. The only open question is who will notice the thresholds as they are drawn, and who will step back long enough to measure them.

Science, spirituality, and the new cosmology

On 27 October 2014, in Vatican City, Pope Francis addressed the Pontifical Academy of Sciences and spoke plainly about the Big Bang and evolution. He did not deliver a physics lecture; he acknowledged scientific accounts of cosmic origins and insisted that such accounts do not, by themselves, abolish the question of creation. The place matters because it is symbolic, and the date matters because it belongs to a period when public narratives still liked their boundaries clean: science over here, theology over there. The outcome was immediate and predictable. Headlines framed it as concession or conflict. But a quieter truth sat behind the noise: both sides were circling the same chalkboard, pointing at the same stubborn feature of reality.

The universe makes sense.

That claim is so familiar we forget its strangeness. We live inside a cosmos that can be described, compressed, and predicted by mathematics; a cosmos whose deep structure yields to measurement; a cosmos that not only exists but *explains itself* to minds made of its own materials. A falling apple becomes an equation. A distant galaxy becomes a spectrum. A whisper of microwave radiation becomes a history of the early universe. The shock is not that humans are clever. The shock is that the world is legible to cleverness.

Legibility is the chapter's organising question, and it is a shared one. Physicists ask it when they wonder why mathematics works so well. Cosmologists ask it when they confront the uncanny fitness of certain constants for complex structures. Philosophers of mind ask it when they try to map experience onto matter without flattening one into the other. Theologians ask it when they resist reducing meaning to mechanism while refusing to deny mechanism's beauty. Different vocabularies, same pressure: why is there order, and why can that order be known?

A five-word sentence belongs here: *Why does anything fit?*

The conversations now gathering around this question are not a single movement, and they are not uniformly friendly. Some are formal, hosted by institutions such as the Pontifical Academy of Sciences, the Vatican Observatory, and academic centres supported by the John Templeton Foundation. Others are informal, carried through books, seminars, and the personal curiosity of scientists who refuse to amputate half their wonder. The phrase "new cosmology" can mislead if taken as a new theory of the universe. The physics remains hard, technical, and contested. What is "new" is the willingness, in some quarters, to admit that explanation has edges—and that those edges touch questions once left to religion or philosophy.

To see the shape of the debate, it helps to begin with a moment that still stings. In 1960, the physicist Eugene Wigner published an essay in *Communications on Pure and Applied Mathematics* titled "The Unreasonable Effectiveness of Mathematics in the Natural Sciences". Wigner did not argue for God, nor did he offer a metaphysical system. He simply named a fact that working scientists often treat as background: mathematics is not merely useful; it is *eerily* apt. Equations invented for internal reasons, guided by elegance or consistency, later describe the behaviour of the physical world with astonishing precision. The behaviour is measurable. Maxwell's equations predict electromagnetic waves; general relativity predicts gravitational lensing and time dilation; quantum mechanics predicts spectral lines and semiconductor behaviour. The metric is predictive accuracy. The implication is philosophical: why should abstract symbols, scratched on paper, grip the fabric of reality so tightly?

One response is deflationary. Mathematics works because it is the language we have constructed to describe regularities; we tailor the tool to the job and then marvel that it fits. This answer has bite. It guards against mysticism. Yet it does not dissolve the puzzle, because it leaves untouched the prior question: why does the world exhibit such stable regularities at all, and why are those regularities compressible into relatively simple mathematical forms rather than requiring an encyclopaedia of ad hoc rules? A universe could have been chaotic in the strong sense—

lawless, or governed by shifting laws, or governed by laws so complex that no finite mind could grasp them. Instead, we find a universe where a small set of principles reaches across scales, from subatomic particles to galaxies, and where approximation works well enough to build technologies that function reliably.

Order is not guaranteed. It is given.

Cosmology sharpens the matter because it forces attention onto constants and initial conditions. The values of parameters such as the cosmological constant, the strengths of fundamental forces, and the masses of elementary particles determine whether stars form, whether chemistry stabilises, whether long-lived structures exist. If these values were wildly different, complex matter might be impossible, and with it the very minds asking questions. This observation is often gathered under the banner of "fine-tuning", and it is a minefield. Some treat it as evidence of design. Others treat it as a selection effect: we observe a universe compatible with observers because incompatible universes cannot be observed.

The anthropic principle, in its modest form, is almost tautological. Yet its implications are not trivial. If our universe appears "just right" for complexity, we must decide what kind of explanation counts as satisfying. One family of proposals appeals to a multiverse: if there are many universes with varying constants, it is not surprising that one supports life. Another family appeals to deeper physical laws that necessitate the values we see, making them less contingent than they appear. A third family—often theological, sometimes philosophical—treats the apparent fit as suggestive of intention. The show–name–explain–implication structure becomes visible here. Show: certain changes in constants would disrupt star formation or chemistry. Name: fine-tuning debates. Explain: competing explanatory frameworks. Implication: cosmology cannot avoid philosophy, because it must decide what counts as an explanation rather than a restatement.

A named, checkable example grounds this. In 1998, two research teams—the Supernova Cosmology Project and the High-Z Supernova Search Team—reported evidence that the expansion of the universe is accelerating, based on observations of distant Type Ia supernovae. The work was conducted across multiple observatories and institutions, with key researchers including Saul Perlmutter, Brian Schmidt, and Adam Riess. The place is not a single room but a network of telescopes and collaborations; the date is fixed. The outcome changed cosmology: it implied the presence of "dark energy" or a cosmological constant-like term, reviving questions about why the vacuum energy density appears so small compared with naïve quantum field theory expectations. This mismatch—the so-called cosmological constant problem—does not merely irritate mathematicians. It presses on intelligibility. Our best theories describe much, yet they do not cohere neatly at this seam. The implication is humility. The cosmos is knowable enough to astonish, and opaque enough to offend our desire for closure.

Humility is not surrender. It is accuracy.

The discussion becomes more charged when consciousness enters. Physics has no equation that yields first-person experience in the way it yields the trajectory of a planet. Neuroscience correlates mental states with brain activity with increasing sophistication, but correlation is not identity, and description is not explanation. Philosophers call this the "hard problem" of consciousness, a phrase popularised by David Chalmers in the 1990s. Whether one accepts the framing or not, the tension remains: how do subjective qualities—pain, colour, longing—arise from physical processes that seem, from the outside, purely quantitative?

Here, again, the chalkboard fills with competing sketches. Some argue that consciousness is an emergent property: when matter reaches a certain complexity, experience appears. Others propose that consciousness is fundamental, a basic feature of reality rather than a late-arriving by-product. Some explore panpsychism, the view that mind-like properties are ubiquitous in some minimal sense. Others lean towards illusionism, arguing that our sense of ineffable inner life is a cognitive construction that misleads us about what exists. These are not fringe distractions; they are serious attempts to keep explanation honest. Each comes with costs. Emergence risks sounding like a label for ignorance. Panpsychism risks diluting consciousness into metaphor. Illusionism risks denying the datum that started the inquiry: we are aware.

This is where theology and spirituality re-enter, not as competing sciences, but as interpretive traditions with long experience of interiority. Religious practice has always treated consciousness as morally significant: attention can be trained; desire can be ordered; the mind can be turned towards truth or away from it. Prayer, meditation, contemplation—across traditions—are not merely opinions; they are disciplined experiments in subjectivity. The modern encounter between science and spirituality becomes fruitful when it resists cheap synthesis and instead allows each domain to expose the other's blind spots. Science can warn spirituality against credulity and fantasy. Spirituality can warn science against mistaking measurement for completeness.

A second named, checkable example illustrates how these conversations become institutional rather than merely personal. In May 2007, in Washington, D.C., the mathematician and cosmologist George Ellis and the theologian and physicist John Polkinghorne participated in public dialogues hosted by academic and philanthropic bodies concerned with science-and-religion engagement, including programmes funded by the John Templeton Foundation. Polkinghorne, ordained in the Church of England after a career in particle physics, became an emblem of a particular kind of intellectual double-citizenship: he refused both the dismissal of physics as spiritually irrelevant and the collapse of theology into vague metaphor. The outcome of such dialogues is not agreement on doctrine. It is the normalisation of a shared question-set: intelligibility, law, contingency, consciousness, moral meaning. That normalisation matters because it changes what educated people consider discussable without embarrassment.

Embarrassment is a surprisingly strong force in the history of ideas.

Still, a credible limitation must be placed on the table. Conversations between cosmology and theology can go astray in predictable ways. The first is the "God of the gaps" temptation: inserting divine action wherever scientific explanation is currently incomplete. This is fragile. Gaps close. Arguments based on ignorance age badly. The second is the mirror-image temptation: treating science as a total metaphysic, as though equations exhaust reality. This is also fragile, because it smuggles a philosophical claim—only the measurable is real—under the cloak of empirical authority. The reconciliation is to keep domains distinct while letting them interrogate each other. Physics can describe how the universe behaves. It does not, by itself, tell us why there is a universe rather than nothing, why laws exist, or why minds should care about truth. Theology can speak about meaning and purpose. It cannot, without self-sabotage, pretend to do astrophysics.

Keeping these boundaries is not timidity. It is discipline.

A further implication follows for moral life. If the universe is intelligible, then truth is not merely a human preference; it is a relationship with the structure of reality. Scientific practice becomes, in this light, a moral practice: it requires honesty, patience, willingness to be corrected, and communal accountability. These are virtues, not only methods. The same is true of mature spirituality when it is at its best. It seeks alignment with what is real, not merely comfort. It trains discernment, not only feeling. The convergence here is subtle but important: both science and spirituality, when healthy, resist the domination of ego.

Ego loves certainty. Reality rarely offers it.

The "new cosmology" conversation also forces a re-reading of older texts. Medieval thinkers such as Thomas Aquinas wrote in a world where cosmos and meaning were assumed to cohere. Modernity fractured that assumption, often with good reason: religious institutions sometimes resisted inquiry, and science sometimes allied itself with reductionism. The emerging conversation does not return us to medieval synthesis, because we now know too much, and our pluralism is too real. Instead, it offers a different possibility: a chastened, plural, inquiry-driven sense that intelligibility itself is a clue worth respecting.

The motif of the chalkboard returns here, not as a classroom cliché but as a symbol of provisionality. Chalk marks are erasable. They invite correction. They are made in public view. The best physics is written that way. The best theology should be, too. Both should be willing to say: this is what we see; this is what we think it means; this is where we might be wrong.

The future of this conversation will not be settled by one argument or one discovery. It will be shaped by institutional choices: what universities fund, what journals publish, what questions are permitted in public without ridicule, what kinds of intellectual courage are rewarded. It will also be shaped by personal choices: whether people can tolerate the discomfort of not knowing while continuing to seek what is true.

The forward test is therefore pointed. When you encounter the universe's intelligibility—whether through a mathematical proof, a telescope image, or the sheer fact of your own awareness—do you treat it as a mere trick of human pattern-seeking, or as a demand placed upon you: to pursue truth with humility, and to ask what kind of cosmos would make such pursuit possible at all?

Ethics written in pattern

In March 2015, in Durban, South Africa, a restorative justice circle convened in a community hall under the auspices of the Phoenix Zululand Restorative Justice Project, a programme associated with the University of KwaZulu-Natal. The case was not symbolic. A young man had stolen from a local shopkeeper. The conventional outcome would have been charge, court, record. Instead, the circle produced a different outcome: a restitution plan agreed by victim, offender, family members, and community representatives, with follow-up meetings scheduled to verify compliance. People did not merely speak; they sat in a particular geometry, faced each other, and committed to repeated return.

The shape mattered.

Moral life is often discussed as a set of rules, virtues, or intentions. Those categories are necessary, but they can obscure something more physical: ethics is also an arrangement. It lives in how we distribute power, proximity, responsibility, and repair. When justice is conceived only as a principle, it remains airy. When it is conceived as a pattern—an ordered relationship between persons, institutions, and consequences—it becomes testable. We can ask whether a system holds, whether it harms, whether it bends towards restoration or towards bitterness.

The central claim is simple and unfashionable: ethics has a geometry.

That geometry appears, first, in the oldest symbol of justice in European legal imagination: the scale. The figure of Lady Justice with balanced scales is not a mere allegory. It expresses a procedural commitment. Weigh evidence. Compare claims. Avoid partiality. Decide proportionately. This is an architecture of judgement, designed to restrain impulse. The behaviour is visible in cross-examination, in evidentiary rules, in sentencing guidelines. The metric is consistency: similar cases receive similar treatment. The implication is legitimacy. A system that cannot show its balancing acts loses moral authority because decisions appear arbitrary.

But scales also conceal. They suggest that justice is principally about equivalence—one harm measured against another, one penalty weighed against an offence. In many cases, equivalence is appropriate. In others, it is morally impoverished. Harm is not always commensurable. A stolen car can be replaced; a death cannot. A scale can "balance" only by simplifying what is morally complex. The limitation is inherent in the symbol. If justice is reduced to balancing, it risks treating moral life as accountancy.

The counter-example is not a rejection of the scale but an enlargement of the picture.

Restorative justice offers a different geometry: the circle. A circle does not rank participants by elevation or distance from authority. It makes faces visible. It frames wrongdoing not only as a breach of rule but as a breach of relationship. This reframing is not sentimental. It changes behaviour. In a circle, the offender must listen to the victim's account. The victim can ask questions that court procedure often excludes. The community can describe the wider damage and the wider support that may be necessary for repair. The metric is not only punishment delivered but restitution completed, relationships stabilised, and reoffending reduced over time. The implication is that justice can be an act of re-weaving, not merely an act of cutting.

A named example grounds the point. In 1989, in New Zealand, the Children, Young Persons, and Their Families Act established family group conferences as a central mechanism of youth justice. The place is fixed; the year is fixed; the outcome is documented: a nationwide shift away from purely court-centred responses towards structured meetings involving the young person, family, victim, and professionals, with agreed plans for repair. The design is not mystical. It is procedural. Yet the procedure is a pattern: a repeatable sequence that uses face-to-face geometry to alter responsibility. New Zealand's model has been studied and adapted internationally, with ongoing debate about effectiveness and implementation, but it remains one of the clearest institutionalisations of circle-based justice in modern law.

The circle, however, is not automatically virtuous. It can be coerced. It can pressure victims into forgiveness. It can mask power differences under the pretence of equality. This is the limitation that must be stated, because it protects the argument from naivety. The reconciliation is to treat restorative geometry as a tool that requires safeguards: informed consent, trained facilitators, and clear boundaries. Geometry influences behaviour; it does not purify motives.

A five-word sentence belongs here: *Patterns decide who gets heard.*

This line is not metaphorical. In any system—court, workplace, family—voice follows structure. Who speaks first matters. Who has time matters. Who can interpret the rules matters. Justice as geometry insists that ethical analysis must include process design, not only moral language. When people complain of unfairness, they are often describing patterned experiences: being interrupted repeatedly, being asked for "proof" others never provide, being penalised for lateness in systems that ignore transport inequity, being assessed by metrics that reward visibility over value.

Legal scales and restorative circles are familiar images. Systems thinking adds a third geometry: the network. Networks do not depict morality as a bilateral dispute but as an ecology of feedback loops. This perspective has become influential in public health, environmental policy, and organisational governance because it captures a hard truth: good intentions can yield harmful outcomes when systems are misaligned. The behaviour is measurable in repeated failures—policies that increase homelessness while aiming to reduce it, policing strategies that increase distrust while aiming to increase safety, welfare reforms that trap people in bureaucracy while claiming to empower them. The metric is not the stated aim but the observable pattern of results.

Systems thinking asks different moral questions. Where do harms accumulate? Which incentives amplify them? Which interventions create unintended consequences? Which feedback loops stabilise injustice?

A named, checkable example illustrates how moral geometry becomes policy. In 1986, the public health scholar

Geoffrey Rose published "Sick individuals and sick populations" in the *International Journal of Epidemiology*, arguing that population-level outcomes often require shifting the entire distribution of risk rather than focusing only on high-risk individuals. Though the article concerns health, its ethical implications are broader. Rose's argument is a geometric one: change the curve, not only the outliers. The implication for justice is clear. If harm is patterned across a population—through housing, education, employment, discrimination—then focusing solely on "bad actors" addresses symptoms while leaving structure intact.

This is where justice meets design most directly. Consider the "scales" of sentencing and the networked realities of inequality. Two people commit the same offence, but one has stable housing, legal representation, and social support; the other does not. The scale may treat them equally in theory, yet the outcomes diverge dramatically. The behaviour is visible in bail decisions, in plea bargains, in reoffending rates. The metric is disparity. The implication is unsettling: formal equality can coexist with patterned injustice.

Ethics written in pattern refuses to accept that as inevitable. It insists that fairness must be evaluated not only at the point of judgement but across the whole system of causes and consequences. This is where some readers become nervous, because the argument can sound like a call for utopian redesign. It is not. It is a call for intellectual honesty:

if outcomes are patterned, then moral responsibility includes pattern diagnosis.

Diagnosis requires a kind of seeing. Earlier chapters explored how minds stitch meaning from noise and how pattern seeking can go astray. Ethical pattern recognition must therefore be cautious. It must resist conspiratorial certainty. It must also resist moral laziness. It should measure, compare, and test.

Measurement does not mean reducing people to data. It means looking for repeatable outcomes that reveal structure.

Here, a credible counter-example keeps the argument grounded. There are moral goods that resist geometrisation. Compassion is not a diagram. Forgiveness cannot be mandated. Love cannot be engineered. Attempts to mechanise virtue often produce brittle systems that punish spontaneity and suppress dissent. The reconciliation is to recognise that ethical geometry is not a complete account of moral life; it is a necessary layer. Patterns can create conditions where virtue is more or less likely to flourish, but they do not replace virtue.

This distinction matters because it protects both sides. It prevents systems thinking from becoming a cold substitute for human judgement. It prevents virtue talk from becoming an excuse to ignore structural harm.

Notice how the motif of the circle returns, now expanded. A circle is not only a restorative practice; it is also a symbol of mutual obligation. When a community decides that repair requires repeated return, it commits to time as well as sentiment. When a court insists on procedural fairness, it commits to sequence as well as verdict. When a policy-maker maps feedback loops, they commit to consequences as well as intentions. These are moral commitments expressed as patterns.

The practical implication is that ethical debates should shift from pure outrage to pattern questions. Instead of asking only "Who is to blame?", we should also ask "What repeats?" Instead of asking only "What do we believe?", we should also ask "What do our systems train us to do?" These questions produce different arguments because they demand evidence. They also demand humility, because patterns are often more complex than our preferred villains.

The forward test, then, is not ideological. It is diagnostic. When you encounter a moral conflict—at work, in community, in public life—do you instinctively reach for the scale, the circle, or the network? Do you weigh and punish, restore and reweave, or redesign the feedback loops that keep harm repeating? And if you choose one geometry, what costs are you willing to bear to keep it from becoming a new form of injustice in disguise?

Living as a conscious motif in the cosmos

On 24 December 1968, as Apollo 8 rounded the far side of the Moon, astronaut William Anders photographed the Earth rising above the lunar horizon. The place was cislunar orbit; the date is fixed; the outcome was cultural as much as scientific. *Earthrise* became one of the most reproduced images in modern history, and its effect was measurable in a shift of public imagination: people began to speak of "Spaceship Earth," to picture the planet as a single, bounded home with a fragile skin. A photograph altered ethics. It changed what seemed thinkable. It did so by compressing vastness into a single frame.

Pattern works like that.

This book has argued—sometimes gently, sometimes with deliberate provocation—that the universe organises meaning through repeatable forms. Not because meaning is reducible to geometry, but because geometry is one of the ways meaning becomes legible. We notice spirals in shells, ratios in music, symmetry in faces and sanctuaries, cycles in calendars and breath. We map and we chant. We build and we repair. In each case, a pattern is not merely found; it is *inhabited*. The final step is therefore not theoretical. It is personal and communal. If pattern is the grammar of attention, then the question becomes: what sentences are you writing with your life?

A five-word sentence belongs here: *Your life repeats, daily, always.*

This is neither comfort nor threat. It is description. The body repeats. The mind repeats. Communities repeat. Workplaces repeat. Even rebellion repeats, if it becomes habitual. The future is not only a horizon; it is an accumulation of small recurrences. To live as a conscious motif in the cosmos is to accept that your choices are not isolated acts but shape-makers. You are laying down a line, and the line will be walked again.

The first claim in this section can be shown without moralising. Watch any person on a normal weekday. They wake at a time that is rarely chosen afresh. They check a device or do not. They speak to someone or do not. They eat in a rhythm that either steadies or agitates. They move through a sequence of thresholds: home to street, desk to meeting, conversation to silence. These movements form a pattern whether they are noticed or not. Naming the reality clarifies the implication: you already have a liturgy. The only question is whether it was designed or drifted into existence.

Drift is a powerful architect.

The second claim is that communities amplify patterns. Individuals can improvise. Communities institutionalise. Shared time makes behaviour repeatable at scale. A family dinner, a weekly gathering, a civic ritual, a workplace schedule: these are not merely logistics. They are moral structures. They decide what is honoured. They decide who is included. They decide what kind of attention is normal. The behaviour is visible in attendance, in lateness tolerated or punished, in who speaks and who remains silent. The metric is retention: whether people return, and whether they return freely. The implication is communal character. A community is the pattern it sustains.

This is why the motif matters. In art, a motif is not the whole work. It is a recurring figure—sometimes subtle, sometimes loud—that gives coherence across complexity. A motif can be beautiful or oppressive. It can invite recognition or induce boredom. It can become a tyrant if it forbids variation. But without motif, a work becomes noise.

The same is true of a human life.

A named example clarifies how conscious motifs can reshape collective behaviour without grand speeches. In 2001, the organisation Alcoholics Anonymous published the fourth edition of *Alcoholics Anonymous*, reflecting continued reliance on the Twelve Steps as a repeated structure for recovery. Whatever one thinks of its theological framing, the method is patterned: meetings at set times, repeated narratives, shared phrases, sponsorship, return. The outcome is visible in a specific behaviour: people who had become unreliable through addiction begin to show up predictably, to repair relationships, to live within a rhythm that supports sobriety. This is not magic. It is geometry in time. The implication is not that one programme fits all, but that repeated structure can convert chaos into intelligible life.

A second example, different in register, shows how motif operates in civic ethics. On 3 September 2020, in Washington, D.C., the National Memorial for Peace and Justice and the Legacy Museum—projects of the Equal Justice Initiative founded by Bryan Stevenson—continued their work of linking historical memory to present responsibility through repeated acts of naming. Visitors encounter rows of suspended steel monuments inscribed with county names and victims of lynching. The design is a pattern of recurrence: names repeat across space, violence repeats across geography, silence repeats across generations. The outcome is behavioural for many visitors: people seek out their counties, contact local officials, initiate soil collection projects, or begin local memorialisation efforts. Memory becomes action through a repeated form. The implication is that ethical awakening often requires design that can hold attention long enough for responsibility to become concrete.

Yet an honest account must face a counter-example. Pattern can be used to anaesthetise conscience. Totalitarian regimes know this. So do exploitative organisations. They impose rituals, slogans, and repeated images that train obedience and extinguish nuance. The limitation is therefore essential: motif is not automatically holy. Repetition can sanctify or stupefy. The reconciliation lies in discernment of *fruit*—visible outcomes over time. Does this pattern increase truthfulness? Does it distribute care? Does it allow confession and correction? Does it permit dissent without retaliation? These are observable behaviours, not private feelings.

A motif that cannot be questioned becomes an idol.

This book has repeatedly returned to thresholds. The threshold is the quiet image that threads the argument because it is where pattern becomes choice. Thresholds are the moments when you can step differently. A doorway. A morning. A screen unlocking. A meeting beginning. A ritual starting. A conflict escalating. A threshold is a hinge between the pattern you have inherited and the pattern you might build.

Living consciously means treating thresholds as sacred opportunities for design.

Design here does not mean aesthetic fussiness. It means deliberate arrangement. It means choosing the rhythm by which you will live, rather than allowing the loudest system around you to choose it for you. In earlier chapters, we examined how pattern comforts and unsettles, how false orders seduce, how digital systems act as pattern priests. This section gathers those threads into a single practical question: what patterns are you consenting to, and what patterns are you composing?

Consent is visible. It appears in the micro-decision to check a feed rather than face silence. It appears in the decision to accept meeting culture that penalises rest. It appears in the decision to tolerate a social circle's cruelty because it is familiar. It appears in the decision to keep returning to a community that refuses repair. These are not abstract moral debates; they are repeated behaviours with predictable outcomes. The metric is time allocation. The implication is formation. You become what you practise.

This is not an accusation. It is a reminder that moral life is not primarily a sequence of heroic moments. It is a cadence.

Cadence is also where hope lives. People often despair because they cannot change everything. Pattern thinking refuses that paralysis. It suggests a different strategy: change what repeats, and let effects accumulate. This is not naïve. Systems resist change. Habits relapse. Communities fracture. Yet patterned change has a distinctive power precisely because it is small enough to be repeated and stable enough to carry meaning across time. When you alter a rhythm, you alter the conditions under which other choices are made.

A person who begins each day with ten minutes of silence is not only adding silence. They are adding a threshold that interrupts compulsion. A community that restores a weekly shared meal is not only eating. It is rehearsing belonging. A workplace that commits to predictable, bounded working hours is not only improving morale. It is redesigning the moral economy of attention.

The danger, again, is rigidity. Conscious pattern must not become control masquerading as virtue. Life requires variation, surprise, and mercy. A healthy motif invites improvisation without losing shape. It holds in storms without becoming a prison.

The cosmos itself models this. Natural patterns repeat, but never as photocopies. No two snowflakes are identical; yet their symmetry is recognisable. Seasons recur, yet each year has its distinct weather. Branching laws yield infinite forms. The implication is almost embarrassingly direct: a holy pattern is not sterile. It is generative.

Generativity is a diagnostic. Patterns that produce life—creativity, repair, courage, clarity—deserve repetition. Patterns that produce decay—fear, concealment, contempt, exhaustion—should be interrupted, however familiar they feel.

The unresolved tension at the end is that consciousness increases responsibility. Once you see your life as motif, innocence becomes harder. You cannot unknow what your patterns do. You cannot pretend that harm is accidental if it is repeated. You cannot claim virtue if your systems reliably injure the vulnerable. At the same time, you cannot excuse despair if small, repeatable changes can be made. The forward test, then, is stark and intimate: when you next cross a threshold—tomorrow morning, in your next conversation, in the next decision to return or to refrain—what pattern will you choose to reinforce, and what will you finally decide to break?

About the Author

Elodie Varkash writes at the meeting point of spirituality, science, and art. For over two decades she has explored how pattern shapes human experience, working alongside contemplative communities, designers, and educators to translate abstract ideas into lived practice. Her work often begins with a simple question: what happens when we treat geometry not as homework, but as prayer in disguise?

Raised among buildings and objects where tiles, arches, and carved motifs quietly told stories, she developed a lifelong fascination with the ways cultures weave belief into form. She draws on historical examples from sacred architecture, manuscript illumination, and traditional crafts, alongside insights from psychology and systems thinking, to show how meaning and structure reinforce one another. Elodie's writing combines clear explanation with a patient respect for mystery, inviting readers who love both rigour and reverence. Her wider mission is to help people redesign the patterns of their days, spaces, and communities so that their outer lives better reflect their deepest convictions.

www.ingramcontent.com/pod-product-compliance
Lightning Source LLC
Chambersburg PA
CBHW031337230926
51848CB00002B/272
9789375368540